D0900289

SOCIAL SCIENCE PERSPECTIVES ON MEDICAL ETHICS

Culture, Illness, and Healing

VOLUME 16

Editors:

MARGARET LOCK

Department of Humanities and Social Studies in Medicine,
McGill University, Montreal, Canada

ALLAN YOUNG

Department of Humanities and Social Studies in Medicine,
McGill University, Montreal, Canada

Editorial Board:

LIZA BERKMAN

Department of Epidemiology, Yale University,
New Haven, Connecticut, U.S.A.

RONALD FRANKENBERG

Centre for Medical and Social Anthropology, University of Keele,
England

ATWOOD D. GAINES

Departments of Anthropology and Psychiatry, Case Western Reserve
University and Medical School, Cleveland, Ohio, U.S.A.

GILBERT LEWIS

Department of Anthropology, University of Cambridge,
England

GANANATH OBEYESEKERE

Department of Anthropology, Princeton University,
Princeton, New Jersey, U.S.A.

ANDREAS ZEMPLÉNI

Laboratoire d'Ethnologie et de Sociologie Comparative,
Université de Paris X, Nanterre, France

SOCIAL SCIENCE PERSPECTIVES ON MEDICAL ETHICS

Edited by

GEORGE WEISZ

Department of Humanities and Social Studies in Medicine,
McGill University, Montreal, Canada

Withdrawn
University of Waterloo

KLUWER ACADEMIC PUBLISHERS

DORDRECHT / BOSTON / LONDON

Library of Congress Cataloging in Publication Data

Social Science perspectives on medical ethics / edited by George
Weisz.
         p.   cm. -- (Culture, illness, and healing ; v. 16)
     Based on a conference held at McGill University in 1988.
     Includes bibliographical references.
     ISBN 0-7923-0566-3
     1. Medical ethics--Congresses. 2. Medical ethics--Social aspects-
  -Congresses.   I. Weisz, George. II. Series.
  R724.S596 1990
  174'.2--dc20
                                                             89-24626

ISBN 0–7923–0566–3

Published by Kluwer Academic Publishers,
P.O. Box 17, 3300 AA Dordrecht, The Netherlands.

Kluwer Academic Publishers incorporates
the publishing programmes of
D. Reidel, Martinus Nijhoff, Dr W. Junk and MTP Press.

Sold and distributed in the U.S.A. and Canada
by Kluwer Academic Publishers,
101 Philip Drive, Norwell, MA 02061, U.S.A.

In all other countries, sold and distributed
by Kluwer Academic Publishers Group,
P.O. Box 322, 3300 AH Dordrecht, The Netherlands.

Printed on acid-free paper

All Rights Reserved
© 1990 by Kluwer Academic Publishers
No part of the material protected by this copyright notice may be reproduced or
utilized in any form or by any means, electronic or mechanical
including photocopying, recording or by any information storage and
retrieval system, without written permission from the copyright owner.

Printed in The Netherlands

TABLE OF CONTENTS

NOTES ON CONTRIBUTORS

Renée C. Fox is Annenberg Professor of the Social Sciences at the University of Pennsylvania. Among her books are (with Judith Swazey) *The Courage to Fail: A Social View of Organ Transplant and Dialysis* (1974), *Essays in Medical Sociology: Journeys into the Field* (1979), and *The Sociology of Medicine: A Participant Observer's View* (1989). She is currently completing a book based on many years of first-hand research in Belgium and Zaire. Mailing Address: Sociology Department, 3718 Locust Walk, Philadelphia, PA, 19104-6299.

Barry Hoffmaster is an Associate Professor in the Department of Philosophy and the Department of Family Medicine at the University of Western Ontario, London, Ontario, N6A 3K7. His main areas of interest are medical ethics, moral philosophy, and philosophy of law. With Ronald Christie he is the author of *Ethical Issues in Family Medicine* (1986). He currently is completing an international study of the ethical decision-making of family doctors in Canada, the United States and Britain.

Christina Honde received her B.A. from the University of California, Berkeley. During twenty years' residence in Japan, Ms. Honde worked as an interpreter, translator, and language teacher, and assisted with research in medical anthropology. She is presently employed as an interpreter, and is working with a Japanese psychiatrist on a translation of his work into English.

Bruce Jennings is Associate for Policy Studies at the Hastings Center, 255 Elm Road, Briarcliff Manor, New York, N.Y., 10510. He has published extensively on topics involving the role of ethics in public policy. Among his recent books are *The Ethics of Legislative Life* (1985) and the anthology (with Daniel Callahan) *Ethics, the Social Sciences, and Policy Analysis* (1983).

Joseph M. Kaufert is a Professor in the Department of Community Health Sciences at the University of Manitoba, 750 Bannantyne Ave., Winnipeg, Manitoba, R3E 0W3. He is a Medical Anthropologist who has published widely, most recently in *Sociology of Health and Illness* and *Social Science and Medicine*. He is currently engaged in ethnomedical research on medical interpreting and renal disease among Native Canadians and sociological research on aging with a disability and impact of life support technology.

Patricia A. Kaufert is Associate Professor in the Department of Community Health Sciences University of Manitoba, 750 Bannantyne Ave., Winnipeg, Manitoba, R3E 0W3. She has published numerous articles about women's health issues, most recently in *Social Science and Medicine, Maturitas* and *Culture, Medicine and Psychiatry*. She is currently investigating childbirth and mid-life among Inuit women.

Richard W. Lieban is Professor Emeritus of Anthropology at the University of Hawaii at Manoa, 2424 Maile Way, Porteus Hall 346, Honolulu, HI, 96822. He is the author of *Cebuano Sorcery* (1967, paperback reprint 1977) and of many articles, most recently in *Culture, Medicine and Psychiatry*.

Charles W. Lidz is Professor of Psychiatry and Sociology and Associate Director for Research Administration of the Center for Medical Ethics at the University of Pittsburgh. His research includes descriptive studies of: informed consent procedures in medical care, how psychiatrists assess risk to others, and the effect of nursing home organization on patient autonomy. His most recent book is *Informed Consent, Legal Theory and Medical Practice* (1987). Mailing Address: Department of Psychiatry, 3811 O'Hara Street, Pittsburg, PA, 15213-2593.

Margaret Lock is Professor of Medical Anthropology at McGill University. She is the author of *East Asian Medicine in Urban Japan* (1980), and editor (with Edward Norbeck) of *Health, Illness and Medical Care in Japan* (1987) and (with Deborah Gordon) of *Biomedicine Examined* (1988). Her research interests are in the anthropology of the body, the cultural construction of life-cycle transitions, and comparative medical systems. Mailing Address: Department of Humanities and Social Studies in Medicine, 3655 Drummond Street, Room 416, Montreal, P.Q., H3G 1Y6.

Edward P. Mulvey is Associate of Child Psychiatry at the University of Pittsburgh School of Medicine, 3811 O'Hara Street, Pittsburg, PA, 15213-2593. His research interests are primarily focused on the use of clinical discretion in social control agencies such as the juvenile court and mental hospitals.

John D. O'Neil is an Assistant Professor of Medical Anthropology in the Department of Community Health Sciences, University of Manitoba, 750 Bannantyne Ave., Winnipeg, Manitoba, R3E 0W3. He has published numerous articles, most recently in *Human Organization* and *Medical Anthropology Quarterly*.

Ian Robinson is Director of the Centre for Research in the Sociology and Anthropology of Medicine and Health at Brunel, The University of West London, Oxbridge, Middlesex, U.K., UB8 3PH. After obtaining degrees in sociology at Nottingham University, England, he has pursued research on a range of issues in medical sociology, focusing particularly on chronic illness. His most recent book is *Multiple Sclerosis* (1988).

David Rothman is Bernard Schoenberg Professor of Social Medicine and Director of the Center for the Study of Society and Medicine at Columbia University, College of Physicians and Surgeons, 630 West 168th Street, New York, N.Y. 10032. Among his books are *The Discovery of the Asylum* (1971), *Conscience and Convenience: The Asylum and its Alternatives in Progressive America* (1980) and (with Sheila Rothman) *The Willowbrook Wars* (1984). He is completing an analysis of the origin and implications of the new regulation of medical decision-making.

Margaret Stacey is Emeritus Professor of Sociology at the University of Warwick, Coventry, U.K., CV4 7AL. She graduated from the London School of Economics in 1943. Among her publications are *Tradition and Change: A Study of Banbury* (1960), *Women, Power, and Politics* (1981) (with Marion Price) which gained the 1982 Fawcett Prize, and *The Sociology of Health and Healing* (1988). She is currently writing up her research on the General Medical Council (to be published by Wiley Medical).

George Weisz is Professor of the History of Medicine at McGill University. He is the author of *The Emergence of Modern Universities in France 1863-1914* (1983), and editor (with Robert Fox) of *The Organization of Science and Technology in France 1808-1914* (1980). He is currently completing a book about the Academy of Medicine and the medical elite in 19th-century France. Mailing Address: Department of Humanities and Social Studies in Medicine, 3655 Drummond Street, Room 416, Montreal, P.Q., H3G 1Y6.

Allan Young wrote this essay while Professor of Anthropology at Case Western Reserve University. He now holds the same position at McGill University. He has conducted ethnomedical research in Ethiopia, Nepal, Israel and the United States. His recent work has been published in *Medical Anthropology Quarterly*, *Culture, Medicine and Psychiatry*, and the *Annual Review of Anthropology*. He is now working on a book on the cultural construction of Posttraumatic Stress Disorder. Mailing Address: Department of Humanities and Social Studies in Medicine, 3655 Drummond Street, Room 416, Montreal, P.Q., H3G 1Y6.

ACKNOWLEDGEMENTS

The editor of any collective volume accumulates a great many intellectual debts. The chief ones, undoubtedly, are to the contributors. Let me here express my gratitude to my fifteen co-authors for producing the contents of this book and, more personally, for the ideas and stimulation which they have provided. I also owe very special thanks to my colleagues in the Department of Humanities and Social Studies in Medicine, Don Bates, Margaret Lock, and Ted Keyserlingk, who were a constant source of ideas and advice and who allowed me to monopolize a disproportionate share of departmental resources while I prepared this volume. Our shared experience of interdisciplinarity has shaped the intellectual vision which gave rise to this book.

I would not have survived the arduous experience of editing an interdisciplinary (and intercontinental) volume without the cheerful and very capable assistance of Elsbeth Heaman. I would like to express my thanks to her and to John Roston and Alexandra Doroschin of McGill's Instructional Communications Centre for producing the printed text with amazing speed. I would also like to thank Miriam Zehavi for typing endless streams of editorial revisions and Harvey Blackman for preparing the index.

This book grew out of a conference held at McGill University in 1988. I would like to take this opportunity to thank all those who participated, particularly my colleagues in the McGill Centre for Medicine, Ethics and Law, for teaching me much of what I know about the subject of this volume.

In preparing that meeting, I received organizational assistance of a very high calibre from Pam Dunk. That this conference took place at all is due in large measure to Dean Richard Cruess of McGill's Faculty of Medicine who intervened at a crucial moment to ensure funding. Even more important, Dean Cruess has over the years, and in spite of an environment of economic austerity, made the McGill Medical Faculty a place where both the social sciences and medical ethics could flourish.

I also gratefully acknowledge the generous financial support for the conference provided by the J.W. McConnell Family Foundation, the Social Sciences and Humanities Research Council of Canada, McGill's Faculty of Graduate Studies and the British Council.

Finally, I would like to express my very special thanks and my love to Zeeva, Talia and Jonathan who put up with my absences and, worse, my frequently preoccupied presence, while helping me get through the more difficult moments in the production of this book.

George Weisz

INTRODUCTION

GEORGE WEISZ

INTRODUCTION

I

Medical or *bio-* ethics has in recent years been a growth industry. Journals, Centers and Associations devoted to the subject proliferate. Medical schools seem increasingly to be filling rare positions in the humanities and social sciences with ethicists. Hardly a day passes without some media scrutiny of one or another ethical dilemma resulting from our new-found ability to transform the natural conditions of life.

Although bioethics is a self-consciously interdisciplinary field, it has not attracted the collaboration of many social scientists. In fact, social scientists who specialize in the study of medicine have in many cases watched its development with a certain ambivalence. No one disputes the significance and often the painfulness of the issues and choices being addressed. But there is something about the *way* these issues are usually handled which seems somehow inappropriate if not wrong-headed to one trained in a discipline like sociology or history. In their analyses of complex situations, ethicists often appear grandly oblivious to the social and cultural context in which these occur, and indeed to empirical referents of any sort. Nor do they seem very conscious of the cultural specificity of many of the values and procedures they utilize when making ethical judgments.

The unease felt by many in the social sciences was given articulate expression in a paper by Renée Fox and Judith Swazey which appeared in 1984. The authors described the medical ethics ("medical morality" in their words) they had been able to study in the course of fieldwork in the People's Republic of China taking care to distinguish these ethics from those characteristic of American bioethics. Against this backdrop, the authors were able to mount a sophisticated critique of American bioethics based on its failure to recognize the social and cultural sources and implications of its own thought. They characterized as culturally specific and intellectually problematic the uncritical emphasis in American bioethics on the individual and his rights (as opposed to the web of human relationships that engender mutual obligations and interdependence), the techniques of rational abstraction that uproot issues from their concrete human reality, and the assumption that these techniques and values have universal applicability.

> In our sociological view, the paradigm of values and beliefs, and of reflections on them, that has developed and been institutionalized in American bioethics is an impoverished and skewed expression of our society's cultural tradition. In a highly intellectualized but essentially fundamentalistic way, it thins out the fullness of that tradition and bends it away from some of the deepest sources of its meaning and vitality (Fox and Swazey 1984: 34).

3

G. Weisz (ed.), Social Science Perspectives on Medical Ethics, 3–15.
© 1990 *Kluwer Academic Publishers. Printed in the Netherlands.*

Ethicists have not been unaware of such issues. Indeed, Fox and Swazey cited a number of critical analyses of bioethics by practitioners like Daniel Callahan. In the years since, there has been a determined effort in some bioethics circles to transcend the prevailing provincialism. Journals in the field publish special issues and supplements focusing on international and comparative perspectives on medical ethics.[1] Many of the articles are by foreign counterparts of American bioethicists and describe parallels to bioethics in other countries. However a recent issue of *The Journal of Medicine and Philosophy* (1988) has at least attempted to explore in greater depth alternative ethical traditions particularly those of Asia. This effort is based on a recognition that there are numerous ethical visions operating in the world. Religions, ideologies, philosophies all generate different and sometimes conflicting responses to problematic issues. In the words of the issue's editor, Robert Veatch,

> Whenever patients standing in one of these alternative religious or secular ethical traditions seek health care from a clinician or health care system within the Hippocratic tradition, there is the potential for a clash between two or more of these traditions. Whenever a health care professional sees himself or herself as standing simultaneously in the Hippocratic tradition and in some other religious or philosophical tradition – as a Catholic or a Jew or secular liberal – there is the potential for a clash of ethical systems within the individual (Veatch 1988: 226).[2]

Apart from the obvious need to take account of most of the planet's population, recent openness to cultural variety and social conflict reflects a growing tendency in some circles to retreat from the notion of universally applicable norms and to embrace, with more or less enthusiasm, the cultural embeddedness of ethical values. This challenge to normative rationality has taken many forms since Joseph Fletcher (1966) first proposed a "situation ethics" not subject to absolute moral rules. Burrell and Hauerwas (1977) have called for greater recognition by ethics of the role of "narratives," the stories which individuals and communities utilize to give meaning to situations and events. Stephen Toulmin (1986) has suggested that the interaction between medicine and ethics has "saved" ethics from its traditional role of relating problematic cases to general theories. In the "case morality" which he advocates the central task is to understand every facet of a particular case.

> In ethics, as elsewhere, the tradition of radical individualism for too long encouraged people to overlook the "mediating structures" and "intermediate institutions" (family, profession, voluntary associations, etc.) which stand between the individual agent and the larger scale context of his actions.... Meanwhile in moral theory, the difference[s] of status (or station) which in practice expose us to different sets of obligations (or duties) were ignored in favor of a theory of justice (or rights) that deliberately concealed these differences behind a "veil of ignorance" (274).

More radical ethical contextualists, represented in this volume by Barry Hoffmaster's essay, argue that there is no "rational" method of morality, that moral decision-making is a matter of "muddling through" problems as they present themselves. As a consequence, the ethicist's principal task is to understand the practice of morality in any situation by locating it in its social and historical context. Contextualism may be part of the "spirit of the times" but one would like to think it reflects real experiences practicing medical ethics among populations that are ethnically and culturally diverse.

Experience clearly has something to do with another sort of retreat from normative ethics. Ethicists increasingly find themselves in forums seeking to formulate public policy. It may well be as Fox and Swazey argue (1984: 34) that this "advisory role to decision-makers has reinforced the cognitive predisposition of bioethics to distil the complexity and uncertainty, the dilemmas and the tragedy out of the situations they analyze." But policy-making committees can also teach important lessons. Those serving on such bodies quickly learn that "muddling through" rather than applying universal principles is the normal mode of operation (Toulmin 1981; Mendeloff 1985). They are confronted in a very direct way by the reality of cultural differences, political and social conflict, and the social construction of consensus. Furthermore, ethicists on such bodies regularly rub shoulders with social scientists of various types. It is hardly an accident, I suspect, that the Hastings Center, a trend-setting institution within American bioethics, published several years ago an excellent collection of essays on the role of the social sciences in policy formulation (Callahan and Jennings 1983).

Social and historical context, competing cultural values, the role of personal narratives, norms related to specific social roles, political and social conflict, these are the stock and trade of social scientists, at least those of a certain bent. One would expect them to have something to say about a medical ethics thus conceived. It is true that social scientists often approach such matters in ways that are different from and uncongenial to ethicists. It is equally true that the various social sciences are experiencing an epistemological disarray that rivals anything faced by bioethics. Still they should, in theory, be capable of making three sorts of contributions to discussions of medical ethics.

First, they can provide ethicists with data, ranging from descriptions of the historical origins of current ethical debates to information about how people in different cultures and at different social levels actually behave in ethically problematic situations. This is probably the least controversial contribution they can make; isolated individuals have done so for years.[3]

Second, the perspectives they utilize may in fact subvert traditional schemas of ethical analysis. Viewing a clinical trial, as Ian Robinson does in this volume, not as an example of the collective welfare which may conflict with the welfare of the individual but as an effort by special interest groups to discredit rival therapies is a profoundly subversive inversion of perspective. Seeing an ethical problem in its broad social context may necessitate its recategorization from the ethical to the political domain. Demonstrating, as Lidz et al. (1984: 315) have done, that the vision of informed consent that is embedded in law and bioethics does not exist in clinical reality, would suggest a need for some rethinking of informed consent doctrine. Such inversions can be both irritating and invigorating for ethical analysis.

Third, subjecting medical ethics and its practitioners (as opposed to issues and dilemmas) to examination and analysis by outsiders may foster the kind of critical self-reflection necessary for the intellectual development of any new field, particularly if it has grown as rapidly as this one.

The social sciences, I should add, have at least as much to gain from this collaboration as does bioethics. Medicine has assumed a cultural importance in modern societies that goes far beyond its ability to make people feel better. Understanding its role has become fundamental to understanding our culture. The development and international diffusion of bioethics is itself a social phenomenon of considerable importance with consequences for our understanding of the entire medical enterprise. Social scientists ignore it at their peril. Furthermore, it would not harm our own efforts at disciplinary self-reflection if ethicists, and particularly the analytic philosophers among them, were to function as critical observers of our working assumptions and intellectual procedures.

These then are some of the reasons for and objectives of this collection of essays. Although several take as their subject the relationship between the social sciences and medical ethics, the majority are *examples* of social science perspectives in action. The volume is not meant to be comprehensive. Readers will notice the absence, due to my own proclivities, of the kinds of applied social science utilized for policy formulation and evaluation. The essays here exemplify another sort of social science, one which seeks to understand the social context and cultural meaning of medical ethics.

II

What do we in fact mean by "medical ethics"? One immediate consequence of adopting social science perspectives is to extend the usage of the term. The papers in this volume refer to medical ethics in at least five different senses, all equally valid.

The most common usage refers to a series of issues which have agitated public opinion during the past decades (and in some cases longer). These include informed consent, human experimentation, the conditions for prolonging or ending life and new reproductive technologies. The list tends to lengthen as new issues like AIDS and the distribution of scarce resources are appropriated. As well as being moral dilemmas, these can also be viewed as complex social, cultural and political processes which emerge and evolve at different times and in different ways from one country to the next.

Another usage of "medical ethics" refers to the network of institutions, commissions and journals within which the agenda of issues is determined and subsequently debated. This institutional network seems to have emerged in the United States during the late 1960s. It was and continues to be dominated by non-physicians, particularly philosophers, theologians and legal scholars. Over the years, the American model of medical ethics (often referred to as "bioethics") has spread widely. It is not, however, the only possible model of institutionalized medical ethics. In European countries like France and Britain, medical ethics grew out of the institutions of the organized medical profession. It is only recently that these institutions have been forced to come more or less to terms with "bioethics." By now, the American model seems to have become the norm so that European observers can perceive their own physician-dominated medical ethics as backward.[4] North American physicians, moreover, have been uncharacteristically reticent to publicly defend traditional physician ethics.[5]

It is not entirely clear why a largely non-medical bioethics emerged first in the U.S. rather than elsewhere. A complete answer would have to take account of the power structures characterizing medicine in different countries. American medicine, for instance was never dominated by a national elite controlling professional deontology and discipline. It may well be that this created a vacuum which allowed outsiders to appropriate medical ethics. Another subject for consideration is the academic market for philosophers in different countries, as well as the predominating intellectual traditions, which may account for the extraordinary affinity of American philosophers for medical ethics. To the extent that bioethics reflects dissatisfaction with contemporary medicine, the possibility in Europe during the 1960s and 1970s of channeling dissension into radical political movements, as well as the greater availability of "alternative" forms of medicine (like homeopathy), may have directed reformist energies on that continent in other directions.

There are a number of other usages which expand the definition of medical ethics even further and thereby allow us to conceptualize issues in new ways. For instance, several authors in this volume understand medical ethics as the cultural values and norms which guide healing-related behavior. In this sense, all societies have some form of medical ethics. As in other domains of human existence, such values and norms may become problematic. A given cultural norm may not provide unambiguous guidelines for action in a specific situation. Several contradictory ethical rules may be appropriate or any single rule may lead to undesired consequences. Finally, individuals and groups may come into conflict because they pursue different ethical goals. To a social scientist, certain forms of behavior may be totally unproblematic within the cultural values of a given society but seem unethical or morally questionable from the point of view of another. The last two situations are increasingly common in our own complicated societies which require different ethnic and cultural groups to co-exist in close proximity.

Healers, health-care workers and researchers may have their own distinct values and norms to guide professional activity. Much of the time such norms are implicit. But occasionally, they are formalized in codes of ethics, books and articles. Such efforts at formalization are often made in moments of crisis and consequently have political and ideological motivations. Much of the historical and sociological work on codes of ethics, in fact, has sought to uncover such social and individual motivations.[6] But precisely because they are designed to serve a variety of purposes, such political documents may not be the best guides to professional ethics as they influence day-to-day activity. Somehow, we need to grasp the implicit norms that actually guide professional behavior.

Finally, "medical ethics" can refer to a particular style of discourse, a way of defining and talking about some aspect of reality. Problems can be conceptualized in various ways, as political, technical, spiritual, for instance. To define a problem as ethical is an intellectual and often social strategy. Among other consequences, it enables social actors engaged in various forms of conflict to tap into the legitimation carried by certain cultural values in our society. Their opponents in such battles must either insist that they too represent ethical principles that are even more critical or redefine the situation in a way that denies ethical legitimacy to competing claims. Defining a problem as "ethical" may also imply that certain types of solutions implemented by particular professionals and institutions are more appropriate than others. One of the crucial tasks of the social scientist is to explain how, from among the nearly infinite number of issues, problems and conflicts which could be defined in ethical terms, some become identified as such while others do not.[7]

Linked to this is the relationship between medical ethics and the political arena. Almost every political ideology appeals to some ethical values and denies ethical value to competing ideologies. Yet the politics of health care is not widely perceived as an issue of medical ethics in contrast to, say, the distribution of scarce health resources which has recently been appropriated by American bioethics. We need to know more about the mechanisms which allow political issues to become ethical issues and ethical issues political. Advocacy groups, the media, institutions of medical ethics expanding their turf, politicians courting popularity or seeking to avoid unpopular decisions, all may be involved in these processes which define the terms in which public issues are discussed.

The different meanings of the term "medical ethics" are in fact overlapping and complementary. They express the various facets of an extraordinarily complex cultural phenomenon. Each is represented to some degree in this volume which is divided into four parts.

The first focuses on dilemmas associated with the clinical encounter. Human experimentation in general, and random trials in particular are usually viewed as ethically problematic. Ian Robinson, in the opening essay, calls into question a basic premise of conventional reasoning on the subject: that random trials represent the "collective interest" which must be reconciled with the interests of the individual patient. After discussing current debates about the scientific validity and limitations of random trials, he focuses on the case of hyperbaric oxygen therapy as a treatment for multiple sclerosis. Robinson argues that trials must be viewed in the context of struggles over power and resources. This does not mean that the results necessarily fail to correspond to the collective interest. But since results are in many cases likely to be inconclusive, the end product may have little effect on clinical practice. This needs to be taken into account in any discussion of the moral dilemmas surrounding clinical trials.

Joseph Kaufert and John O'Neil next provide an equally subversive analysis of informed consent. Arguing that traditional legal and ethical discussions which seek to apply abstract rules are inadequate for understanding actual situations, the authors insist that any discussion of the subject must take account of the different explanatory models for the understanding of disease and therapy utilized by health-care workers and patients. They must also take account of power relationships that are highly unbalanced. Both these matters are particularly visible in the case examined here where patients are Natives confronting the Canadian health care system. Here both the discrepancy between explanatory models and the unequal distribution of power are so great that native-language interpreters are needed to mediate between the parties. The complex way they perform this task is the subject of several detailed case studies.

Allan Young demonstrates the fruitfulness of utilizing an expanded definition of medical ethics. In his usage "ethics" are the moral values according to which individuals organize their behavior. He goes on to analyze the ethical conflicts and dilemmas which arise in a veterans hospital specializing in the care of posttraumatic stress disorder (PTSD). Occasionally, the ethical character of conflictual encounters is recognized by the actors. Most often it is not because the inequalities of power which different individuals wield in this milieu encourage a purely medical definition of issues. Young particularly emphasizes the role of etiological narratives about the origins of PTSD in shaping actors' understanding of what is taking place.

Finally, Charles Lidz and Edward Mulvey use both quantitative and qualitative techniques to measure the effect of financial pressures on the admitting practices of a psychiatric emergency unit. They come to the conclusion that professional norms of beneficence continue to have considerable influence on doctors' decisions and lead them to resist many sorts of financial pressure. While a single case is obviously not representative, it suggests that administrative efforts to influence clinical behavior through financial constraints are likely to encounter formidable difficulties.

Part II is devoted to medical ethics in the public arena. Margaret Lock shows how the question of organ transplant has become a controversial public issue in Japan. Advocates of organ transplants have had to utilize a strategy of "going public" in seeking to overcome various forms of cultural resistance to the practice; chief among these is a widespread unwillingness to accept brain death as the termination of life. This strategy has also been made necessary by the Japanese belief in achieving change through public consensus. Lock's paper illustrates the vital role played by cultural beliefs and values in matters of medical ethics. It suggests that there may be far more variation in the way such matters are handled in different countries than is usually thought to be the case. It also raises some troubling questions about the lack of public debate in the West about such issues.

Patricia Kaufert describes from the perspective of a participant observer the political battle which broke out in Canada over the licensing of the contraceptive drug Depo-Provera. It is a cautionary tale about shoddy experimentation and economic exploitation, among other things. But for the purposes of this book, its importance lies rather in the conflict of discourses which took place, and in the effort of women's groups to transform the terms of debate. Rather than accepting that this was a technical decision to be taken by experts, opponents of licensing redefined it as an issue of medical ethics in which such matters as informed consent operating at the collective and political levels were at stake. The paper shows in important ways how a sociological and feminist perspective can expand and transform traditional views of medical ethics.

The papers in Part III examine institutions of medical ethics. Those by George Weisz and Margaret Stacey look at the ethics generated by medical elites in France and Britain. The former focuses on the first international conference devoted to medical ethics which took place in Paris in 1955. Weisz argues that medical ethics emerged in France at this time in response to three developments: the creation of a national body with disciplinary power over French medicine; concerns about human experimentation that grew out the Nazi medical atrocities; and fears that the physician's therapeutic autonomy was being threatened by the generalization of the national health insurance system. The Congress allowed the French medical elite to popularize an individualistic vision of medical ethics that has continued to be influential in that country.

Stacey examines the evolving guidelines which Britain's General Medical Council has been providing to practitioners during the past decade. In a careful review of guidance publications, she shows that the Council faces the problem of reconciling its role of protecting professional interests with its role of protecting the public. In the past, there has been a tendency to be concerned with intra-professional issues, but since the early 1980s the Council has devoted more attention to protecting the public. Her essay analyzes the different types of pressure which have led to changes in the Council's guidance of practitioners.

The scene then shifts to this side of the Atlantic in David Rothman's study of the role of human experimentation in the emergence of the contemporary bioethics movement. More than the geographical location has changed, however. The nascent movement which Rothman describes is very different in character from the physician-dominated ethics found in Europe. According to Rothman, this movement grew out of the growing public controversy over human experimentation which erupted in the early 1960s. This was the first issue to bring an array of lawyers, philosophers, clergymen and elected officials into the world of medicine. Rothman chronicles the process which led gradually to the imposition of outside authority on certain traditional domains of physician responsibility.

The essay by Renée Fox bridges this section on the institutions and ideology of medical ethics and Part IV which explores some of the theoretical issues involved in the study of medical ethics by social scientists. Bringing to bear first-hand experience as a pioneering participant observer in the sociology of medical ethics, she first describes incisively three stages in the evolution of American bioethics since the 1960s. She goes on to analyze the philosophical and ideological presuppositions which underlie the field and then attempts to explain the lack of dialogue between ethicists and sociologists. Fox particularly emphasizes the differing ethos and intellectual procedures characteristic of each discipline.

Richard Lieban next examines the silence of medical anthropology on the subject of ethics. He attributes this silence to a number of different factors including the dominant cultural relativism of the field. Using material from his own field notes as well as examples from recent anthropological writings as illustrations, he suggests ways in which the discipline can contribute to our knowledge of a broadly defined medical ethics that takes account of cultural variation.

Two ethicists examine in turn the nature of possible ties between medical ethics and social science research. Barry Hoffmaster provides us with a vigorous philosophical critique of the positivist position which currently dominates medical ethics. This culminates in a defense of ethical contextualism which involves locating a moral problem in its social and historical context. Such an approach, he argues, depends on the findings of social science research for its effectiveness. Hoffmaster then discusses several recent and exemplary social science publications of this type.

Lest we become unjustifiably optimistic about the possibilities for interdisciplinary consensus, Bruce Jennings examines recent writing on neo-natal intensive care units and shows how different are the perceptions of ethnographers from those of ethicists. In analyzing the reasons for such differences, he denies that ethicists are concerned with the "ought" whereas ethnographers are interested in what "is." Rather both observe functioning units on the basis of particular values and views of how things ought to be. It is the ways of observing together with views about possibilities for change which differ to produce such divergent accounts.

The book ends with two bibliographies. The first lists books that either exemplify or are relevant to social science research on medical ethics. (Although articles have not been included, I have included several dissertations and special thematic issues of journals.[8]) A second bibliography lists very selectively some of the most important books in the field of medical ethics. It is intended for non-specialists who would like to pursue an interest in the subject.

III

The papers in this volume were originally presented at a small conference held at McGill University in June 1988. Although they were substantially revised for publication, the book as a whole would have been inconceivable without the meeting. For that reason and because the experience may have some relevance to future efforts in this area, I should like to describe it briefly.

In planning the conference, I tried to assemble a group of scholars who were actually doing (as opposed to talking about) research on medical ethics utilizing social science perspectives. To counteract the prevailing North American provincialism, I also made an effort to include individuals who either studied or had experience with medical ethics in other countries. Both tasks proved to be more difficult than I had imagined. I quickly learned that there is no recognizable community of social scientists working in the field and that the published literature is still relatively sparse. (It occurs to me that if medical ethicists make so little use of relevant research in the social sciences it may be in part because they cannot find it.) I had consequently to rely on personal recommendations in order to identify scholars who might have something substantial to contribute. By the time I was forced to commit closure (which prevented me, unfortunately, from inviting at least a half-dozen more individuals who came late to my attention) I had assembled a very eclectic group of about thirty individuals. Sociologists predominated, reflecting both the expertise of my informants and the fact that medical sociology has been, relatively speaking, more conscious of issues of medical ethics than other social sciences. I also made an effort to invite a number of practicing ethicists to participate on the assumption that the social sciences have no monopoly on insight.

Nineteen papers were pre-circulated and then discussed over the course of a three-day meeting. These discussions took place, it seemed to me, in an atmosphere of considerable intensity reflecting a common awareness that a new research community might be in the making. At the same time, the disparity of backgrounds and the variety of approaches taken sometimes made communication painful.

Differences of disciplinary perspective emerged time and again during the course of the conference. As one would expect, there was some tension between ethicists and social scientists based on differences of technique and aims. However, neither ethicists nor social scientists are part of monolithic communities. Each is fragmented along disciplinary and theoretical lines. Consequently, the nature of the meeting between ethics and social science is very much dependent on the sort of subgroups involved. Many of the participants, for instance, were struck by the apparent compatibility of a certain strand of medical sociology based on ethnographic research methods and certain tendencies in medical ethics.

My choice of medical ethicists, it should be emphasized, was not necessarily very representative of the field. Nearly all were sympathetic to the effort to inform medical ethics with the results of social science research. By the close of the meeting, some ethicists were asking sociologists like Renée Fox and Charles Bosk for guidance in improving the ethnographic skills of ethicists. This prompted Fox to remind ethicists that their aims differed substantially from those of sociologists and that they had to cultivate their own brand of practical ethnographic enquiry.

Not everyone participated in the friendly exchanges between ethnographers and ethicists. The anthropologists at the meeting remained surprisingly silent during these particular discussions. I assumed at the time that this silence resulted from doubts about the relevance of ethnographic fieldwork in non-western societies to western ethical dilemmas. It has since been suggested to me that some anthropologists found these discussions impossibly naive given current debates within their own discipline about the epistemological and even ethical status of ethnographic enquiry.

A more serious challenge to the ethnographic perspective was a tendency among certain sociologists and social historians to view both ethical dilemmas and the responses of medical institutions as embedded in if not determined by social structures and political processes, notably those related to distinctions of class and power. As it happened, the most persistent voices at our meeting on behalf of this type of analysis were British. It may well be that European sociologists and historians are more likely than their American counterparts to choose such macro-sociological perspectives. But many North American scholars share these views.

Dialogue between ethnographers and ethicists will not be without its tensions. The former will be more likely to seek to understand reality in all its complexity whereas the latter are increasingly being asked to solve immediate problems. (There is of course a long tradition of more abstract moral philosophy but my impression is that the recent institutional growth of bioethics largely reflects the expansion of applied ethics.) Still, the realities they are examining are similar: individuals acting within institutions. Scholars examining the socio-ideological context of medical ethics, however, focus on a rather different level of reality, one whose relevance to solving the problem at hand is not immediately apparent. If they have a message, it is all too often that individual action to resolve a problem is pointless unless social arrangements are transformed. They may suggest that the problem being confronted, the response of medical institutions *and* the interventions of the ethicist are all determined by social and ideological forces. This is hardly comforting to the participants involved. My guess is that this form of social science research will never have more than a marginal impact on the actual exercise of medical ethics. But it is nonetheless vital to a full understanding of the enterprise.

Some participants at the McGill conference felt that certain key issues had not been adequately discussed. Several, for instance, complained that social and ideological issues had been slighted. One sociologist expressed astonishment that the critical role of religion had not been addressed, as it was not being addressed by medical ethics in general. She wondered whether, in our secular age, religion had now become an unmentionable taboo. As conference organizer, I was disappointed in my inability to locate more scholars able to analyze the institutions and personnel of medical ethics as well as their relationship to the political process and the media.

Overall, however, the areas covered proved very substantial. If disciplinary division and conflict were not completely overcome (one historian told me that, after three days, he had concluded that there was *less* to medical ethics than meets the eye), participants were for the most part pleasantly surprised by the degree of cross-disciplinary dialogue that had in fact occurred. There was widespread consensus that this could and should provide a basis for future research.

This volume of essays is one consequence of that agreement. One can only hope that the McGill meeting will turn out to have been the first in a series of conferences and collective volumes permitting a growing community of scholars to develop and exchange views while stretching existing disciplinary boundaries. If that turns out to be the case, we will be on our way to achieving a deeper understanding of the role of medical ethics in different societies and cultures.

ACKNOWLEDGEMENTS

For comments on an earlier draft of this paper, I am grateful to Allan Young, Renée Fox, Don Bates, Margaret Lock, Ted Keyserlingk, Benjy Freedman, Judith Miller, Ian Robinson, Joe Kaufert and Ronnie Frankenberg.

NOTES

1. For instance, the *Hastings Center Report* had special international supplements in 1987 and 1988. Also see *Theoretical Medicine* 1988.
2. Despite my sympathy for the general thrust of this quote, I must as a historian point out that Veatch's identification of ethical values within medical institutions with "the Hippocratic tradition" is simplistic and, in important ways, misleading.
3. For a discussion of sociological work of relevance to bioethics see Fox 1989. More generally see Bibliography I in this volume.
4. Good examples of such reasoning are Isambert 1986 and Sass 1988.
5. An exception is a largely philosophical critique by a philosopher and a physician (Clements and Sider 1983) of the dominance in bioethics of the principle of autonomy. In defending traditional medical paternalism, the paper takes a position that is widespread in Europe.
6. Among works in this "demystifying" mode are Unschuld 1979; Berlant 1975 (chapter 3); Waddington 1975; Naylor 1982-83.
7. It is of course common for social scientists to *engage* in this form of discourse rather than to analyze it. Some individuals see their role as social scientists of medical ethics as a form of "whistle-blowing" in which they brand as "unethical" policies or behavior in the health care field with which they disagree.
8. Unfortunately, it proved unmanageable to include individual articles in this bibliography. To make up partially for this, I have included some books in which there is only a single chapter or essay of relevance to this topic.

REFERENCES

Berlant, Jeffrey L.
    1975   Profession and Monopoly: A Study of Medicine in the United States and Great Britain. Berkeley: University of California Press.
Burrell, David and Stanley Hauerwas
    1977   Commentary: Rationality, the Normative and the Narrative in the Philosophy of Morals. *In* H. Tristram Engelhardt and Daniel Callahan (eds.) Knowledge, Value and Belief (The

Foundations of Ethics and Its Relationship to Science. Vol. II). Hastings-on-Hudson, N.Y.: The Hastings Center. Pp. 111-152.

Callahan, Daniel and Bruce Jennings
1983   Ethics, the Social Sciences, and Policy Analysis. New York and London: Plenum Press.

Clements, Colleen D. and Richard C. Sider
1983   Medical Ethics' Assault Upon Medical Values. JAMA 250: 2011-2015.

Fletcher, Joseph F.
1966   Situation Ethics: The New Morality. Philadelphia: Westminster Press.

Fox, Renée C.
1989   The Sociology of Medicine: A Participant Observer's View. Englewood Cliffs, N.J.: Prentice-Hall.

Fox, Renée C. and Judith P. Swazey
1984   Medical Morality is Not Bioethics – Medical Ethics in China and the United States. Perspectives in Biology and Medicine 27: 337-360.

Hastings Center Report
1987   Biomedical Ethics: A Multinational View 17: 1-36.
1988   Special Supplement: International Perspectives on Biomedical Ethics 18, 4: 1-31.

Isambert, François-André
1986   Révolution biologique ou réveil éthique? In Éthique et biologie (Cahiers S.T.S.). Paris: Centre National de la Recherche Scientifique. Pp. 9-41.

Lidz, Charles W. et al.
1984   Informed Consent: A Study of Decisionmaking in Psychiatry. New York: The Guilford Press.

Mendeloff, John
1985   Politics and Bioethical Commissions: "Muddling Through" and the "Slippery Slope." Journal of Health Politics, Policy and Law 10: 81-92.

Naylor, C.D.
1982-83 The CMA's First Code of Ethics: Medical Morality or Borrowed Ideology? Journal of Canadian Studies 17: 20-32.

Sass, Hans-Martin
1988   Biomedical Ethics in the Federal Republic of Germany. Theoretical Medicine 9: 287-298.

Theoretical Medicine
1988   Medical Ethics in Europe. D. Thomasma (ed.) 9, 2.

Toulmin, Stephen
1981   The Tyranny of Principles. Hastings Center Report 11, 6: 31-39.
1986   How Medicine Saved the Life of Ethics. In Joseph P. DeMarco and Richard M. Fox (eds) New Directions in Ethics: The Challenge of Applied Ethics. London: Routledge and Kegan Paul. Pp. 265-281.

Unschuld, Paul U.
1979   Medical Ethics in Imperial China: A Study in Historical Anthropology. Berkeley: University of California Press.

Veatch, Robert M.
1988   Comparative Medical Ethics: An Introduction. The Journal of Medicine and Philosophy 13: 225-230.

Waddington, Ivan
1975   The Development of Medical Ethics – A Sociological Analysis. Medical History 19: 36-51.

PART I

CLINICAL ENCOUNTERS OF THE ETHICAL KIND

IAN ROBINSON

CLINICAL TRIALS AND THE COLLECTIVE ETHIC: THE CASE OF HYPERBARIC OXYGEN THERAPY AND THE TREATMENT OF MULTIPLE SCLEROSIS

The ethical basis of medical experimentation on human subjects is an issue at the core of bioethics. In such experimentation the balance between individual and collective rights, and in particular the balance between physicians' therapeutic obligation to their patients – the individual ethic (Clayton 1982), and their duty to further scientific knowledge for the collective good – the collective ethic (Lellouch and Schwartz 1971), has become an issue of major importance. Much of the debate about this issue has now been focused on the ethical status and role of the clinical trial in modern scientific medicine.

However it is the central contention of this paper that the considerable effort vested in elucidating the ethical dilemmas and problems in the principles and practice of clinical trials has resulted in a partial and particular view of their ethical status. Specifically, what constitutes the "collective ethic" has been treated as self evident, and has been little explicated. The analysis that follows suggests that the ethical implications of clinical trials are more complex, and embedded in a broader range of social and political forces than is usually assumed.

In the forty years since Bradford Hill's randomised controlled trial of the effects of streptomycin on pulmonary tuberculosis (MRC 1948), the growth of medical research in general, and the growth in the number and size of randomised controlled trials in particular has been remarkable. Such trials have rapidly assumed a fundamental role in the validation of medical knowledge (Johnson and Johnson 1977; Friedman, Furberg and DeMets 1980). Their role in such validation has been extended, at least in principle, beyond therapies located in formal clinical practice into the general domain of health care (Ditchley Foundation 1980: 134). Furthermore, a very considerable proportion of national medically related financial, professional and organizational resources are now devoted to such trials (Kennedy and Bennett 1981).

THE IDEA OF THE RANDOMLY CONTROLLED TRIAL

The potency of the idea of the randomly controlled trial seems to lie at a conceptual level in its assumed scientific purity; therefore by definition in some accounts, it is ethically impregnable in its unique and impartial access to the measurement of effective therapy (Ditchley Foundation 1980; Baum 1982). In this view, by eliminating to the maximum possible extent extraneous factors which might contaminate the examination of the efficacy of the variable under study (a treatment regime), and by the statistical capacity to assess the extent of the remaining biases, the relative efficacy of the regime can be established through these procedures. The randomised controlled trial thus offers the very

G. Weisz (ed.), Social Science Perspectives on Medical Ethics, 19–39.
© 1990 Kluwer Academic Publishers. Printed in the Netherlands.

best means to judge therapeutic efficacy; it is an ethical good of a substantial kind. Clinical medicine is therefore served by a mechanism which produces standardised data on treatment effectiveness – allowing the physician a more informed (and hence more ethically satisfactory) basis for judgement. Further, medical science is served by the same mechanism which in addition allows attention to be focused on the nature of cause and effects.

Such trials are thus perceived to stand at a critical point between the scientific generation of therapeutic knowledge and its clinical application (Johnson and Johnson 1977; Vere 1983). They are capable not only of acting as a channel between scientific possibilities and validated therapeutic interventions, but are also capable of serving as a means through which other domains of knowledge about health and medicine can, indeed must be sifted to separate the therapeutically beneficial, from the therapeutically benign and harmful. Without such a mechanism all medically relevant knowledge is deemed to remain in a state of unrequited potential – a state in which its application remains a random and chance affair full of unknown and unknowable dangers.

In principle, therefore, the therapeutic landscape over which randomised controlled trials could hold sway is vast. Any intervention which is or might be clinically deployed is potentially the object of attention, as is any intervention relating to health which is, or might be deployed by others – for example nurses and therapists (even when they are not subject to direct clinical control), alternative practitioners of any kind, folk practitioners, patients themselves and their families. Even if the specific focus of such trials is limited to drug-based interventions, as the Ditchley Foundation Conference implied (1980), elastic definitions of what constitutes a drug allow very considerable interpretative latitude in evaluation through clinical trials (Hansen and Launso 1987a, 1987b). The ethical evaluation of such potent and wide ranging instruments therefore deserves especially careful scrutiny.

THE DEBATE ABOUT THE ETHICAL BASIS OF RANDOMISED CLINICAL TRIALS

Despite the speed with which randomised controlled trials have become incorporated into medical knowledge and practice, there have always been caveats expressed about their role. The two major areas of concern initially appeared to be: first, the extent to which the results of such trials would interfere with the autonomy of physicians' clinical judgement; and second, the extent to which the necessarily stringent research constraints in such trials would inexorably prejudice the participating physician's basic therapeutic obligation to each patient. Such caveats coming almost entirely from within the medical profession itself, have been seen by the proponents of trials as the vestiges of an unscientific – and potentially dangerous and unethical – vision of medical practice (Baum 1982; Vere 1983). These concerns, and the response to them, represent part of what Freidson (1970) describes as the tension between the clinician healer and the clinician scientist centred, in this case, on an apparent major reconfiguration of the basis of medical knowledge. However the ethical virtues of the randomised controlled trial as the final arbiter of scientific knowledge appear to have become more fully and formally accepted, at least among medical researchers. Cochrane's (1972) powerful plea for the use of clinical trials as the main mechanism through which medical therapies should be tested has additionally reinforced their position as an instrument for establishing the collective good. Nonetheless others have continued to debate the status of clinical trials by referring to problems inherent in their structure and operation, or by noting the considerable discrepancy between trial findings and clinical practice. The most extreme view in this respect is that randomised controlled trials are intrinsically unscientific (Burkhardt and Kienle 1983), or intrinsically unethical (Marquis 1986), or both (Burkhardt and Kienle 1983). In large measure, the two areas of debate have been conducted in tandem, with arguments at their starkest indicating that scientific imperatives in the trial setting present insoluble ethical problems, or vice versa. There have been assertions that technical solutions can be found to many of the previously intractable problems (Armitage 1982). However at the heart of this debate still lies the putative incompatibility between the "individual ethic" and the "collective ethic" (Schaffner 1986: 299). The question can be phrased thus: is it possible to have a randomised controlled trial in which the obligations of the clinician scientist to the collectivity do not ethically undermine the obligations of the clinician healer to individual patients participating in the trial? In brief, is it possible to reconcile the scientist's duty to all present and future patients with the healer's duty to his current patient?

A series of debates have been conducted around this question relating to when a trial can ethically be held (Schaffner 1986); the nature and effects of informed consent (Baum 1986; Macara 1986; *Journal of Medical Ethics* 1982; Kirby 1983; Kopelman 1986; *Lancet* 1984; Levine 1987; Prout 1981); the effect of randomisation on the therapeutic obligation to individuals (Gifford 1986; Schaffner 1986; Waldenstrom 1983); the extent and timing of disclosure of information about a trial to participants (Kadane 1986; Van Eys 1982; Von Eickstedt 1983); and the basis on which a trial should be continued or stopped (Schaffner 1986).

For some writers, in the light of these debates, the fundamental answer to the question posed above is clearly a negative one (Marquis 1986). For others there are devices by which the conflict might be reduced although it is still a major difficulty (Gifford 1986). For yet others new forms of adaptive or sequential trial design clearly have the potential to overcome most of the difficulties (Armitage 1982), a possibility made more likely by the application of the logic of Bayesian principles (Kadane 1986). The general issue has been placed in especially sharp focus by West German legislation of 1979 which appears to place such a heavy emphasis on the therapeutic obligation of physicians to their individual patients that randomised controlled trials may, in one view, be impossible to carry out effectively (Burkhardt and Kienle 1983). This interpretation, although disputed (Hasskarl 1981), has nonetheless raised additional concerns about the ethical viability of clinical trials. Baum has more recently argued that such a situation is occurring even without legislative intervention where

> We have reached a point where almost everyone would approve an unreliably controlled treatment comparison, but many would consider formal reliably controlled comparison unethical ... (1986: 912).

However there are those who still assert that there is no ethical conflict at all (Ditchley Foundation 1980). As individuals are part of the collectivity they gain in equal measure from the undertaking of the trial – therefore the beneficial results of trials are ethically indivisible. It is but a short step from this point to argue that there is an ethical duty on the individual to participate in such a trial not because the collectivity will benefit but because it is in their own best therapeutic interests. Thus the individual and collective ethics are perceived as coincident.

This view of the congruence of individual and collective benefits of clinical trials seems considerably at variance with the way in which clinical practice appears to operate. The extent to which clinicians fail to adhere to the practices that might be legitimately expected in implementing trial findings (Chalmers 1982), raises a range of questions about organization of trials themselves, as well as about their subsequent collective benefits.

The apparently marginal effects of trial results on clinical practice may be because, as some commentators have argued, in many cases only a modest amount of therapeutic information can be drawn from the findings of such trials (Arpillange and Dion 1986). By and large any extra increment of benefit established is likely to be statistically (and clinically) small over "the previous best alternative." Findings are more frequently negative in the sense of establishing that therapies are of no proven benefit compared to alternative kinds of intervention and, more rarely, in establishing statistically harmful benefits compared to those alternatives (or, as appropriate, placebos). From the point of view of the scientific application of trial findings the latter two categories can be deemed therapeutically equivalent – that is "no benefit" or "harm" implies the same clinical decision. Indeed following this line of argument to recommend intervention with an unnecessary (i.e. unbeneficial) remedy is in itself either actually or potentially harmful, and hence results in an ethically dubious clinical action.

However this view is problematic for the clinician healer faced with therapies of broadly equivalent, but limited efficacy, in which the distinction between therapies of "no benefit" and those which "harm" is a material one. Chronic and life threatening conditions, where by definition proven and effective therapies are generally unavailable, raise this issue in a dramatic way. The nature of the formal criteria utilised in trial designs, in relation to their limited focus on a narrow range of (primarily) biomedical factors, casts doubt on the easy applicability of their mainly negative results to such situations (Gilmour 1977; Simnett 1982; Greer 1984). Increasing clinical interest in the general quality of life of those who are chronically ill (Teeling-Smith 1987), together with the realisation of the signifi-cance of psychosocial factors in chronic or progressive conditions (Hunt, McEwan and McKenna 1986), has reinforced concern over the relevance of the biomedically centred findings of many trials to the clinical management of such conditions (Robinson 1987). In these cases, biomedically "harmless" therapies with no (scientifically) proven benefit may be clinically employed if they exhibit a measure of psychosocial potency (Robinson 1988: Chap. 5). Moreover there may not only be a discrepancy between trial findings and individual clinical strategies, but in this context the collective virtues of the results are themselves rendered problematic.

This discussion points up the potentially difficult relationship between trial results and clinical practice. One view is that such results, embedded as they are in scientific assessments, "set standards to which the patient and the physician can aspire ..." (Ditchley Foundation 1980: 131). In this view the randomised controlled trial is both a necessary and sufficient means to establish the extent of the collective value of a particular therapy. Variations from the practical endorsement of this value, whatever it may be, are treated as evidence of the remediable failings of practitioners and patients. For others, the discrepancy between the strictly controlled world of the trial and the more open and complex world of clinical practice is such as to indicate the need for major changes in trial procedure before the results of the latter can be confidently extrapolated to, and expected in such practice (Hansen and Launso 1988). Thus the failure of practitioners and patients to endorse therapeutic values, as established through trials, is perceived as evidence of narrow trial objectives and problematic protocols.

This continuing debate about the ethical and scientific status of clinical trials appears to have been mainly conducted within a relatively narrow paradigm by clinician healers or medical scientists themselves. More recently the active participation of bioethicists has broadened the debate. However it does appear that many of the controversies over the ethical status of clinical trials have failed to address the broader issue of the genesis of the debate itself.

EXAMINING THE COLLECTIVE CONSEQUENCES OF RANDOMISED CONTROLLED TRIALS

Both Barnes (1979) in his general analysis of the increasing attention paid to ethical issues in relation to research, and Berlant (1975) in his more particular study of the centrality of ethical codes in the professional practice of medicine, point to the linkage between social and organizational change and the shifting focus of moral concerns. Further they each note that the ethical status of these concerns largely depends on the way that they are incorporated into professional practice. In turn, professional standing and power is reinforced by the association with explicit moral objectives. In this kind of analysis, randomised controlled trials assume no special or privileged position by virtue of the attribution of

unique scientific or ethical integrity to them, for that integrity – and any
associated collective good – is inferred by particular social and professional
interests at a particular time, and does not stand independently of those interests.
   Burkhardt and Kienle, for instance, quote Meir as follows:

> [the question of continuing or stopping a trial] ... is, in fact, a political problem, I see
> no sensible way of coping with it outside a political framework. The question, really, is
> what sort of society we wish to live in (Meir 1979).

They then go on to note that

> controlled trials [are] a "social challenge." It is asocial to obscure or minimise the
> problems involved; they need to be faced (1983: 83).

However such a broad challenge to the objective scientific status of clinical trials
uncomfortably displaces traditional debates. One response may be to resist this
unwelcome intrusion of additional issues. Vere reasserts the traditional agenda
based on the respective roles of healer and scientist, as well as pointing to the
social virtue (and the inherent logic) of the scientific modality and its implications for the patient's role. He redirects the argument about the political context
of trials thus:

> Do people wish to remain diseased, or do they wish to be relieved; that is the political
> problem. If they do wish their sufferings eased then they face the logical problem, that
> the only way ahead, whether by clinical trials or by other kinds of experiments, is by
> paying the price – suffering – of acquiring new knowledge by orderly methods.
> Clinical assessments are not the answer ... (Vere 1983: 88).

This response does not appear to deal with a number of fundamental questions
raised by Meir and Burkhardt and Kienle. In particular it fails to acknowledge the
problems inherent in channelling the diversity of ideas about the nature of
disease, possible therapeutic responses to it, as well as the wide range of
individual and social priorities in relation to health, through the single mechanism
of the randomised controlled trial. Furthermore, even if it is assumed for the sake
of argument that such trials are intrinsically asocial and apolitical mechanisms, to
what extent can their selective use, the resources devoted to them, and the
professional, and other interests associated with them be said to operate in a
socially unproblematic way? Or more particularly in relation to the issue which
underpins this analysis, to what extent does their use ensure the effective
operation of a beneficial collective ethic?
   The question of the proper domain of scientific medicine and its territorial
boundaries, as well as the area of knowledge under its control, is an issue posed
by Vere's comments. The increasing importance of community, especially

preventive medicine and more pliable positive, rather than negative definitions of health emphasize the relevance of this issue. Disease-based medicine constitutes a far more modest – although still large – area over which the controlled trial holds scientific sway. However the manner in which medical interests and concerns may follow extended and elastic notions of "health" and the factors which influence it, indicates the prospect of a very considerable growth in the domain in which the trial can persuasively be applied.

A further issue relates not only to the extent of the domain whose knowledge base is potentially capable of arbitration by the controlled trial, but to the nature of the domain. The dominant site of such trials has been the laboratory, hospital or clinic. In addition to being the location of those with the technical expertise necessary to undertake such trials and of necessary resources and equipment, such sites afford the maximum opportunity to dislocate research subjects from their natural environment. This dislocation, part of the process of trial control, may remove, suppress or render invisible the complexity of individual or social objectives, personal strategies and goals which messily complicate ordinary life. Although this is precisely the object of a technique which seeks to determine, identify and isolate particular factors relevant to therapeutic success, results from trials thus derived can only be reinserted into a lived-in social world with grave difficulty (Hansen and Launso 1988). It may be this special difficulty which has led to the substantial discrepancy between the concerted scientific endorsement of clinical trials and everyday medical practices in which trial results often appear to be deployed in a manifestly unscientific manner. This discrepancy may be further compounded by the range of professional and financial interests which surround the running of trials and dissemination of results. The demonstration of the collective benefit of individual clinical trials does not appear to occur spontaneously or effortlessly. It is a highly organised process with both professional and commercial reputations dependent not only on the proper running of trials, but also on the subsequent public management of findings stemming from them.

This latter point suggests that the circumstances in which the trial is undertaken, the ways in which financial interests may condition the therapies that are tested, the professional interests involved in the running and outcome of trials, the extent and kind of regulatory mechanisms which operate, and the broader impact of trials on health and health policy, are all matters of social, political and ethical interest. All involve the allocation or use of resources ostensibly for the public good through the medium of the trial itself.

There are of course considerable commercial interests involved in the outcome of many trials. This is particularly true of the pharmaceutical companies whose role, it has been argued, has a distorting effect not only on the kind of therapeutic products tested, but on the structure and operation of clinical trials themselves (Hansen and Launso 1987b). Individual professional careers, as well as the status awarded to medical sub-specialities may depend to a large degree on the nature of involvement with clinical trials (Barber 1981). The overtly political context in which clinical trials operate is well expressed by Ingelfinger whose analysis is that

> it has to be a matter of the whole public being committed to the principle of clinical trials ... if the voters are [so] impressed they will influence their government (1982: 279).

He goes on to argue for the importance of mass political mobilisation and lobbying to acquire the funds for all the necessary trials. Such a fusion of scientific judgement and political action suggests that careful scrutiny needs to be given both to the nature of the collective benefits of this procedure, and to the (collective) costs that may be thus incurred.

The increasingly important role of legally enforceable regulatory mechanisms also affects the nature of trial processes in balancing a range of different and sometimes competing interests (Ditchley Foundation Report 1980). Thus the dynamic process which is associated with establishing and running trials may not be a totally disinterested and altruistic one. This is not necessarily to say that such a dynamic process has invariably unethical consequences for the collectivity (whatever the participants' original motivations), but that those motivations bring together a chemistry of values and interests which raise the possibility of unethical outcomes.

It appears to be assumed in many of the debates about the outcomes of clinical trials that the "collective ethic" emerges automatically from the operation of clinically based scientific endeavour. This suggests that those involved in clinical research may be characterised as having broadly identical behaviours and beliefs about the consequences of their actions, or that the "hidden hand" of science produces a similar effect. This does not seem to be a satisfactory account of the collective consequences of clinical trials, nor does it address issues raised by the interplay of professional and commercial interests. The complexity of the issues raised by the idea of the collective ethic as a self evident and beneficial outcome of clinical trials is well illustrated by a case study which will now be presented. It indicates the significant role of social factors and professional interests in determining the way trials themselves are generated. The case study also illustrates ways in which trial findings may have a range of effects which cast doubt on an easy assumption of collective benefit.

THE CASE OF HYPERBARIC OXYGEN THERAPY AND MULTIPLE SCLEROSIS

Hyperbaric oxygen therapy has always been a medical therapy marginal to mainstream medical science (Behnke 1977). This continuing marginal status in itself appears to have conditioned the context in which randomised controlled trials were crucial in its later assessment and formal scientific dismissal as a therapy of choice in multiple sclerosis. A brief account of the historical background of the therapy is therefore important. The analysis broadly follows that of Jacobson, Morsch and Rendall-Baker (1965).

The therapy has its origins in diving experiments involving raised atmospheric pressure carried out from the time of the early Greeks. The first medical use of air at raised atmospheric pressure has been documented as occurring in the mid 17th Century. However it was not until two hundred years later that the administration of air, and much later the adminstration of oxygen under increased atmospheric pressure was deemed to be beneficial to health. The prime initiators of many of these early developments were entrepreneurial physicians who created hyperbaric health spas involving compressed air rather than pure oxygen. Such spas were very successful financially and spread all over Europe but declined in the late 19th Century, largely it appears, due to the changing professional practice of medicine and the lessening of claims which could legitimately be made about the effects of the therapy. Use of hyperbaric therapy with oxygen increased again at various points in the first half of the 20th Century, most notably in the large practice of Orval Cunningham who built a remarkable five storey hyperbaric hospital in Cleveland in 1928 to treat a wide range of conditions. However Cunningham was subject to investigation, then damning indictment by the American Medical Association for his scientifically unsubstantiated claims. His practice collapsed, and the hospital closed shortly afterwards, due it appears not to the actions of the A.M.A. but to the effects of the Great Depression on the financial viability of his enterprise.

Since that time there have remained a small number of hyperbaric chambers in the United States, the majority devoted to more conventionally acceptable medical work in diving and space medicine. Indeed an interesting and parallel development to some of the more unconventional uses and users of hyperbaric oxygen (HBO) has been the formation of the Undersea Medical Association, and later the Undersea Medical Society. These organizations were specifically set up to rescue hyperbaric medicine from its marginal, indeed disreputable, status and to regularise its clinical use within a scientifically acceptable framework. However, the unconventional history and the nature of HBO made it likely that any new clinical application of the therapy, which is administered to patients in a pressurised chamber, would be subject to especially severe scientific scrutiny. More specifically, this scrutiny was further conditioned and focused by the following factors.

First, almost all potential applications of hyperbaric therapy outside the narrow domains of diving and space medicine would be likely to intrude on the existing diagnostic and therapeutic procedures of more established medical sub-disciplines. Second, the technology needed for the administration of the therapy, and the therapy itself, was relatively simple and in principle widely available; in consequence it was not likely to prove commercially exploitable on a monopoly basis through the acquisition of unique property rights or patents. Third, the therapy in various forms had already been associated with marginal medical practitioners seeking to treat a wide range of conditions and, just as importantly, had been associated with lay initiatives and endorsement. Fourth, the therapy – oxygen under high pressure – appeared to be a generic, rather than a highly targeted remedy. Its action seemed non-specific at a time of fine diagnostic distinctions and appropriately focused therapies. Fifth, the basic technology for the therapy was anchored in work and activity outside the medical domain altogether, making it more difficult to control solely through medical means.

The potential therapeutic use of HBO in relation to multiple sclerosis appeared to be first publicly recognised from work in the Florida hyperbaric facility of Neubauer, a private physician specialising in hyperbaric medicine, (although there was an earlier positive but little noticed study from Czechoslovakia by Boschetty and Cernoch 1970). Neubauer reported that a patient treated for osteomyelitis (a scientifically validated use) in 1974 appeared to experience a major remission in her multiple sclerosis (Neubauer, Schnell and Goldberg 1979). From that point he began to treat other patients with multiple sclerosis, and in effect set up a large open study on a continuing series of patients using their subjective responses to evaluate the therapy. These responses were overwhelmingly favorable in that up to 90% of patients reported some improvement (Neubauer 1980). The extent of this rate of improvement – which in some of Neubauer's findings appeared to be sustained over long periods – was remarkable in view of the absence of any scientifically validated therapy with a long term effect on the disease process. Through both informal patient networks and increasing publicity about his work in the period up to 1983, Neubauer's hyperbaric facility attracted worldwide attention and numerous patients with multiple sclerosis, further adding to the open series of those reporting beneficial results.

In addition to Neubauer, other physicians working in hyperbaric medicine became interested in the potential of the therapy for multiple sclerosis. In the United States, more hyperbaric facilities began to treat people with the disease. Some were already established centres, others were set up by entrepreneurs who employed the necessary qualified medical personnel. One of the prime proponents of the therapy in Britain was James, a specialist in diving medicine in Scotland, who began using hyperbaric chambers to treat patients. Even more importantly, he developed a theory about the aetiology of multiple sclerosis radically at variance with conventional ideas; in his view the disease had a vascular rather than neurological, virological or immunological cause (James 1982). These developments were in turn followed by, and accelerated through the work of ARMS (Action for Research into Multiple Sclerosis), a British self-help group which began to consider establishing HBO therapy centres outside the formal health care system after contact with Neubauer. It received support from James and other hyperbaric physicians.

ARMS had been founded in 1974 by a group of patients and their relatives who were dissatisfied with the progress and direction of research into the condition.

Further, they were dissatisfied with the lack of patient involvement in decision-making within the larger and longer established British Multiple Sclerosis Society, whose research and medical policies were set and guided by neurological expertise. Unlike the Multiple Sclerosis Society, ARMS' work was focused on active, quick and direct involvement in the management of the condition even, if necessary, where this was at variance with conventional medical opinion. This pattern of a radical but small disease-based charity confronting a larger and more established rival is not uncommon amongst the voluntary organizations involved in health care issues in Britain (Robinson D. and Henry 1977). What is unusual in this case is the longevity of the smaller organization.

Before 1983 neither ARMS nor HBO therapy was considered a serious threat to established medical ideas, or to the Multiple Sclerosis Society. This is not surprising in view of the many unconventional remedies reputed at any one time to have a therapeutic role in relation to multiple sclerosis. Over 200 such remedies have been the subject of critical scientific documentation by the International Federation of Multiple Sclerosis Societies (1982).

However a number of studies during 1982 and 1983 substantially elevated the scientific and medical status of HBO and led to a more serious consideration of its therapeutic role. A junior neurologist and his colleagues presented a paper to the 7th Annual Conference on the Clinical Applications of Hyperbaric Medicine which appeared to show a dramatic series of improvements in patients with multiple sclerosis in a double blind randomised controlled trial. But publication of this paper was delayed for a year before it appeared with a series of caveats in the *New England Journal of Medicine* (Fischer, Marks and Reich 1983). In addition, in 1983 a report on an open trial was published by a British neurologist indicating similarly positive results (Davidson 1983). These studies following earlier reports were then associated with a major series of developments.

ARMS began to build up its infrastructure of local groups using the open and the controlled trial results among other information as a basis for fund raising for the provision of local hyperbaric centres. This development occurred with considerable speed such that in the two years between 1984 and 1986 over 50 local HBO centres were established; membership of ARMS doubled to over 8,000 people and the organization's financial resources rose fourfold. Moreover this development began to have substantial organizational and financial effects on the British Multiple Sclerosis Society (BMSS). In the limited market for members with the disease, recruitment to the BMSS declined, fundraising was affected and, in a number of cases, senior local officials of the BMSS were openly advocating the use of HBO and employing organizational resources to support the therapy (Robinson 1988: 122). In the United States further local initiatives were undertaken to increase the availability of HBO; ARMS in America was established as an organization supporting the growth and availability of the therapy. Senior neurologists who had maintained a considerable distance from this new therapy found that their position was being undermined by three developments. First, a randomised controlled double blind trial – at the apex of evaluative techniques – had apparently proved the therapy beneficial (Fischer, Marks and Reich 1983). Second, an apparently viable theoretical explanation of this beneficial effect had been developed which was opposed to conventional neurological views of the aetiology or management of the condition (James 1982). Third, patients were experimenting with the therapy in ARMS HBO centers in increasing numbers outside neurological control.

Hyperbaric physicians were not only encouraged by the developments to pursue the treatment of multiple sclerosis, but also to do research into other therapeutic uses of HBO. Thus the possibility of playing a central role in the management of a major neurological condition gave a professional impetus to a group of hitherto medically marginal practitioners.

The development and subsequent validation of the therapy through a randomly controlled trial appeared simultaneously to: undermine almost all established therapeutic strategies and many aetiological views; challenge dominant neurological perspectives on the disease; strengthen the professional position of a group of medically marginal practitioners; and give a further impetus to lay involvement in the management of the condition.

In response to this situation established neurologists and national Multiple Sclerosis Societies reacted in four ways. First they began to carefully dissect the Fischer trial through a meticulous examination of the protocol, the procedure and the explanatory model which supported the results. Second they raised doubts about safety in the administration or prolonged use of the therapy. Third they supported new randomised controlled trials to challenge the results of the Fischer trial. Fourth they sought to ensure that as few patients as possible undertook the therapy pending the results of further trials.

In Britain two randomised controlled trials of HBO therapy were established in 1984 funded together with other trials elsewhere by the Multiple Sclerosis Society. The Society actively advised all those with the disease not to participate in the therapy, stressing its dangers and the absence of general neurological support for its use (Robinson 1988: 122).

In anticipation of, or fear that the other trials would fail to meet the highest standards of scientific impartiality, ARMS set up its own trial in 1985. This trial was conducted in so far as possible with scientific rigour in the knowledge that its procedure and results would be severely scrutinised. Other less rigorous trials were also being undertaken as open, or case-controlled longitudinal studies. Many were further series of open cases treated in various HBO centres worldwide (Neubauer 1983). Most of the studies did not include formal neurological assessments of the patients' status and were thus not considered by neurologists to accurately reflect therapeutic effects (Bates 1986). Such criticisms were exacerbated by a substantial reliance on subjective assessments of efficacy.

The first major report of one of the British Multiple Sclerosis Society trials was published in February 1985 (Barnes et al. 1985). It noted that

Because our results were at variance with previous reports we have decided that they be published at an early stage rather than wait a further twelve months for longer term results (297).

Later the authors wrote:

We conclude ... that hyperbaric oxygen is unlikely to have a role in the management of the patient with multiple sclerosis (300).

Another major report of a Multiple Sclerosis Society trial was published one year later (Wiles et al. 1986). It came to almost identical conclusions, indicating that the only benefit from HBO appeared to be some minor subjective improvements. A range of other randomised controlled trials came to similar conclusions. (Confavreux et al. 1986; Lhermitte et al. 1986; Harpur et al. 1986).

The ARMS crossover trial was reported at a Symposium in September 1986 and published in 1987 (Worthington et al. 1987). Its general conclusions were similar to those of Barnes et al. (1985), and Wiles et al. (1986), although more subjective improvements were noted and a limited range of positive changes were recorded. Other open trials, and case controlled studies with substantial subjective assessments continued to record major improvements, far more modest than Neubauer's original study had indicated, but far more congruent with Fischer's trial results (see Webster et al. 1988).

Given the very substantial agreement between a series of randomised controlled studies on the negligible objectively measured benefits of HBO for patients with multiple sclerosis, it might be thought that these results would be problematic for: those patients who still registered subjective improvements with the therapy; physicians from a range of medically marginal disciplines who still believed that the therapy was, or could be made effective; and ARMS, and other bodies elsewhere whose rapid expansion had been based on the use of HBO for the management of the disease. However it is important to distinguish between the debate about the significance and meaning of the trials and the extent to which therapeutic practices may have been transformed as a result of them.

HBO has not been incorporated into the modest range of therapies employed by neurologists in their management of patients with multiple sclerosis. Moreover there is only limited evidence that such an incorporation would have been unproblematic even if trial results had been marginally more encouraging because of the considerable discrepancy between the explanatory frameworks and clinical practices of hyperbaricists and neurologists (see the extended analysis in Nicholson and McLaughlin 1988). In this particular intraprofessional conflict, neurologists, perhaps because of their high medical status and their substantial and continuing ability to control access to their occupational market, seem to have been set on a "rigid" and rejectionist, rather than an "imperialistic" and incorporative course of action, to use Gritzer and Arluke's terms (1985). The direction of neurological practice and research has continued to be largely antagonistic to the vascular theories associated with hyperbaric medicine (Bates 1986), theories of which might have challenged neurological control of the disease entity. It is worth noting in this respect that the hyperbaric facility of Fischer and his colleagues was closed in 1985 with no further research being undertaken on the effects of HBO on neurological conditions. Thus whilst one of the more obvious collective effects of the trials may have been to scientifically undermine substantive therapeutic claims for HBO, a further effect has been to reinforce the status and knowledge base of neurology as a medical sub-discipline.

Nonetheless such a reinforcement following the results of recent trials may be of a more partial nature than it first appears. It has already been noted that there is often a considerable discrepancy between trial results and clinical practice. Clinical autonomy amongst other factors, provides a setting in which doctors may engage in procedures that are at considerable variance with trial findings. In this case, lay control over the means of delivery of HBO and the active support of some hyperbaric physicians allowed the use of the therapy to proceed with minimal regard for either the concerns of neurologists or trial results.

All those ARMS centres at which HBO therapy was provided during the period of major expansion continue to offer it to patients (ARMS 1989). If fewer people now appear to be using the facilities it is because of other factors at least as significant as the consequences of the trial results. In particular, to maintain full use of the ARMS HBO Centres requires more patients than are currently being diagnosed because of the present oversupply of the therapy at a large number of centres set up in the initial enthusiasm for the therapy. This problem is compounded by a treatment regime which is based on intensive sessions followed by more occasional "follow-up" treatments.[1] Conversely, negative trial results themselves have been dismissed by supporters of HBO who criticize the trial protocols or practices and especially their failure to fully take into account the subjective dimensions of changing health status (Webster et al. 1988). Overall it thus appears that logistic and practical, rather than scientific factors may be the most critical ones in determining the future use of HBO for multiple sclerosis. Hyperbaric specialists originally involved with ARMS and the HBO programme are still actively monitoring and researching the technique (Webster et al. 1988), and ARMS itself still incorporates the therapy as part of its strategy for the management of the disease.

Although there are those within ARMS who, supporting HBO therapy, wish to continue the debate about trial findings, the use of the therapy has diminished far less than might be expected if everyday practice had been directly extrapolated from the results of the trials. It appears that the weight of trial evidence itself has not influenced practices in a major way, at least in Britain. The vindication of neurological expertise in relation to multiple sclerosis appears to have been largely confined to the sub-discipline itself, and to those bodies such as the national Multiple Sclerosis Societies whose research policies have been influenced, if not actually determined by neurologists. As in other areas of health care it seems that a substantial difference exists between formal research evidence, and clinical and lay practice (McEwan, Martini and Wilkins 1983). There is little in this case study to suggest that trial findings have significantly increased the congruence between research evidence and practice.

DISCUSSION

This account can be treated as an almost classic example of the triumph of scientific medicine over the misuse of an inappropriate therapy for a serious neurological condition. At least in principle, many future patients may have been saved by the trial results from undergoing a therapy with a known lack of objective benefit; in short they have been saved from a therapy which it would have been unethical to administer. Current patients have been saved further inappropriate and costly action. An account by Bates, a neurologist, summarising the scientific history of the therapy up to 1986 takes precisely this line of argument, although he wonders in view of his assessment

> why the proponents of hyperbaric oxygenation should be so insistent upon their claims that the treatment is of benefit to their patients (Bates 1986: 535).

An analysis of the context within which the therapy arose, was supported and was evaluated would go some way to meet this point which seems ethically partial. To a large degree, controversy over the therapy, and its individual and collective benefits, has hinged on the validity of trial procedures. In particular, such controversy has focused on whether conventional clinical trials are capable of assessing the complexities of the therapeutic effects of HBO on a fluctuating but serious chronic condition, and especially whether, in the available trial assessments, sufficient credence has been given to subjective improvements noted by patients. In this context the results of trials in a condition like multiple sclerosis can often be viewed as ambiguous. Given the considerable range of disease parameters that might be affected, the varied pattern of natural remissions and relapses, the difficulty of standardising objective measurement criteria (Silberberg 1984), and the increasing emphasis on the importance of quality of life with such diseases (Robinson 1987), the scope for multiple interpretations of trial data is substantial. In addition, as Arpillange and Dion (1986) note, the majority of clinical trials even when they are statistically successful on conventional scientific criteria add only marginally to the control of conditions to which they relate. Consequently in clinical practice, the impact of the therapeutic difference identified by statistically successful compared to statistically unsuccessful trials may be modest, even if it is fully implemented. In the particular case of chronic disease such as multiple sclerosis this factor, combined with difficulties of trial measurement and assessment, further clouds the evaluation of trial findings. Therefore, although HBO may, on the weight of scientific evidence from clinical trials, be considered to be no better than the previous best alternative, that alternative realistically is no treatment. To equate "lack of benefit" with "harm" in the sense discussed earlier in this paper seems problematic unless either side effects of such therapies are taken into account (on which criterion HBO is more benign than many other therapies under trial for multiple sclerosis (Robinson 1987), or such non-medical factors as patients' time or resources are considered. If the latter factors are included, other ethical issues are raised about the extent to which patients (citizens?) are allowed the autonomy to deploy their own resources in directions they themselves wish to choose.

The protection of patients with multiple sclerosis from malign (or at least unbeneficial) therapies and from their own potentially misguided actions – thereby enhancing the collective good – seems to have been only one component in the use and interpretation of HBO trials. There were plainly other issues involved for many parties to the debate. The research policy and role of national Multiple Sclerosis Societies was under threat, as was the professional position of neurology in relation to the control of patients with multiple sclerosis. For ARMS and for specialists in hyperbaric medicine there was a major opportunity for institutional growth and enhanced professional status. Some of these factors appeared to influence how the various trials were evaluated. Not all randomised trials were considered to have been of equal merit – even if there was no clear evidence that major aspects of the trial protocol had been broken or poorly executed. It seems that *who* ran the trial as much as *how* the trial was run was a critical factor in an assessment of its validity.

In summary, whilst in principle the ethical status of trials can stand independently of the motivations and interests of those involved in their organization, funding or running, the insertion of competing interests into what has been argued to be an inherently ambiguous and problematic procedure, further complicates the evaluation of their collective benefits.

In retrospect the short-term effect of this very substantial ripple in the lengthy attempt to find an effective therapy for multiple sclerosis has been to consolidate the formal medical authority of neurologists in the clinical management of the condition, and to endorse the research policies pursued by national Multiple Sclerosis Societies largely under neurologists' direction. However, although the theories and the role of hyperbaric specialists in relation to multiple sclerosis have apparently been placed in jeopardy, as have ARMS' policies towards research and therapy, in practice there has been little effect on the strategies pursued by either party. Because much of the therapeutic momentum in relation to HBO has always been outside the control of neurologists, a dual system of managing the disease is operating, one system managed by neurologists, and the other managed by other professional interests, voluntary organizations and patients themselves. In this latter system HBO continues to play a large role at least in Britain. If the role of HBO declines, it appears likely to be largely for reasons other than the negative results of randomised controlled trials.

CONCLUSION

In the introduction to this paper it was argued that the randomised controlled trial is a recent feature of scientific and medical analysis. Its current pre-eminence in the domain of formal health care needs no restating. The fact that the technique, and other associated procedures raise issues of ethical and scientific substance is undoubted. The extent to which randomised controlled trials, as currently organised in relation to chronic disease, are an effective medium through which therapeutic efficacy can be definitively and unambiguously tested has been questioned. The difficulties involved in incorporating a broader range of patient-based psychosocial subjective assessments which complement more limited biomedical assessments in such trials, together with the marginal increments of therapeutic knowledge which trials often generate, suggest that their collective benefits need more critical analysis. Further, the social context in which controlled trials operate is as problematic as their more specific ethical and scientific context. Indeed in considering the social context of trials, further ethical issues are uncovered about the nature and extent of their value. The use of trials and the application of their results pose particular questions about their relationship to health care. Arguments about their collective value generally neither explicate the theoretical basis on which that value is calculated nor analyse the relationship between everyday medical practice and trial results. In the case study above, the collective effects of trials are demonstrated to be complex. They may be related to scientific arbitration between different therapies; related to the fortunes and status of professional groups or specialities; related to commercial success or failure; related to establishing or re-establishing boundaries of commercial expertise and competence in health care. They may also have a range of as yet undocumented consequences on the actions of patients and others. They may also have very limited effects on everyday health care, an issue of concern to those who believe unequivocally in their collective benefits (Cochrane 1972).

Further detailed studies are needed of the conditions and circumstances in which randomised controlled trials are established; of their funders; of their scientific and medical organisers; of the broad consequences of their operation and their results for patients and society at large. Such studies may allow us to take a more considered view of how the collective ethic is constituted through the principles and practice of clinical trials.

NOTES

1. On conservative calculations, ARMS' 50 centres with HBO require an annual additional intake of some 15,000 patients in order to allow them to operate at full capacity using the ARMS HBO treatment regime. Whilst annual incidence figures of the disease in Britain are a matter of some debate, it appears that between 3-4,000 would be a reasonable estimate (Matthews et al. 1985).

REFERENCES

Armitage, P.
1982 The Role of Randomisation in Clinical Trials Statistics. Medicine 1: 345-352.

ARMS
1989 Handbook of ARMS Therapy Centres. Stansted, U.K.: Action for Research into Multiple Sclerosis.

Arpillange, P. and Dion, S.
1986 The Ethical Approach to Randomisation in Oncology Drugs. Experimental Clinical Research 12: 99-103.

Barber, B.
1981 Biomedical Research as a Social Process. *In* Wechsler, H., Lamont-Havers, R.W. and Cahill, G.F. (eds.) The Social Context of Medical Research. Cambridge, Mass.: Ballinger. Pp. 127-142.

Barnes, J.A.
1979 Who Should Know What? Social Science, Privacy and Ethics. Harmondsworth: Penguin.

Barnes, M.P. et al.
1985 Hyperbaric Oxygen and Multiple Sclerosis: Short Term Results of a Placebo-Controlled Double Blind Trial. Lancet 1: 297-300.

Barnes, M.P. et al.
1987 Hyperbaric Oxygen and Multiple Sclerosis: Final Results of a Placebo-Controlled, Double Blind Trial. Journal of Neurology, Neurosurgery and Psychiatary 50: 1402-1406.

Bates, D.
1986 Hyperbaric Oxygen Therapy in Multiple Sclerosis. Journal of the Royal Society of Medicine 79: 535-537.

Baum, M.
1982 The Ethics of Clinical Trials and Informed Consent. Experientia (supplement). 41: 300-308.
1983 Commentary. Journal of Medical Ethics 9: 92-93.
1986 Do We Need Informed Consent? Lancet 2: 911-912.

Behnke, A.R.
1977 A Brief History of Hyperbaric Medicine. *In* Davis, J.C. and Hunt, T.K. (eds.) Hyperbaric Oxygen Therapy. Bethesda: Undersea Medical Society. Pp. 3-10.

Berlant, J.L.
1975 Profession and Monopoly. Berkeley: University of California Press.

Boschetty, V. and Cernoch, J.
1970 *Aplikace Kysliku za Pretlaku Nekterych Neurologickych Anemocneni*. Bratislav Lek Lesty 53: 298-302.

Burkhardt, R. and Kienle, G.
1983 The Practical and Ethical Defects of Surgical Randomised Prospective Trials. Journal of Medical Ethics 9: 90-92.

Chalmers, T.C.
1982 Randomised Clinical Trials and the Consumers. *In* Tygstrup, N., Lachin, J.M. and Juhl, E. (eds.) The Randomised Clinical Trial and Therapeutic Decisions. New York: Marcel Dekker. Pp. 257-263.

Clayton, D.
1982 Ethically Optimised Designs. British Journal of Clinical Pharmacology 13: 469-480.

Cochrane, A.L.
1972 Effectiveness and Efficiency: Random Reflections on Health Services. London: Nuffield Provincial Hospitals Trust.

Confavreux, C. et al.
1986 Hyperbaric Oxygen Therapy in Multiple Sclerosis: A Double-Blind Randomised Placebo-Controlled Study. Presse Médicale 15: 1319-1322.

Davidson, D.L.W.
1983 Hyperbaric Oxygen Therapy in the Treatment of Multiple Sclerosis. Stansted, U.K.: Action for Research into Multiple Sclerosis Education Service.

Ditchley Foundation Report
1980 The Scientific and Ethical Basis of the Clinical Evaluation of Medicines. European Journal of Clinical Pharmacology 18: 129-134.

Fischer, B.H., Marks, M. and Reich, T.
1983 Hyperbaric Oxygen Treatment of Multiple Sclerosis. The New England Journal of Medicine 308: 181-186.

Freidson, E.
1970 Profession of Medicine: A Study of the Sociology of Applied Knowledge. New York: Dodds Mead.

Friedman, L.M., Furberg, C.D. and DeMets, D.L.
1980 Fundamentals of Clinical Trials. Boston: John Wright.

Gifford, F.
1986 The Conflict Between Randomised Clinical Trials and the Therapeutic Obligation. Journal of Medicine and Philosophy 11: 347-366.

Gilmour, D.
1977 Psychological Factors in Clinical Trials. In Johnson, F.N. and Johnson, S. (eds.) Clinical Trials. Oxford: Blackwell. Pp. 147-161.

Greer, S.
1984 The Psychological Dimension in Cancer Treatment. Social Science and Medicine 18: 345-349.

Gritzer, B. and Arluke, A.
1985 The Making of Rehabilitation. Berkeley: University of California Press.

Hansen, E.H. and Launso, L.
1987a Clinical Trials: Towards a Different Model. Journal de pharmacie clinique 6: 67-74.
1987b Development, Use and Evaluation of Drugs: The Dominating Technology in the Health Care System. Social Science and Medicine 25: 65-73.
1988 Is the Controlled Clinical Trial Sufficient as a Drug Technology Assessment? Working Paper, Royal Danish School of Pharmacy, Copenhagen.

Harpur, G.D.
1986 Hyperbaric Oxygen Therapy in Chronic Stable Multiple Sclerosis: Doubleblind Study. Neurology 36: 988-991.

Hasskarl, H.W.
1981 Legal Problems of Controlled Clinical Trials. Controlled Clinical Trials 1: 401-409.

Hunt, S., McEwan, J. and McKenna, S.P.
1986 Measuring Health Status. London: Croom Helm.

Ingelfinger, F.J.
1982 Randomised Controlled Trials and the Public. In Tygstrup, N., Lachin, J.M. and Juhl, E. (eds.) The Randomised Clinical Trial and Therapeutic Decisions. New York: Marcel Dekker. Pp. 277-281.

International Federation of Multiple Sclerosis Societies
1982 Therapeutic Claims in Multiple Sclerosis. New York: International Federation of Multiple Sclerosis Societies.

Jacobson, J.H., Morsch, J.H.C. and Rendall-Baker, L.
1965 The Historical Perspective on Hyperbaric Oxygen. Annals of the New York Academy of Sciences 117: 651-670.

James, P.B.
    1982   Evidence for Sub-acute Fat Embolism as a Cause of Multiple Sclerosis. Lancet 1: 380-386.
Johnson, F.N. and Johnson, S. (eds.)
    1977   Clinical Trials. Oxford: Blackwell.
Journal of Medical Ethics (Editorial)
    1982   Research, Consent, Distress and Truth. Journal of Medical Ethics 8: 59-60.
Kadane, J.B.
    1986   Progress Towards a More Ethical Method for Clinical Trials. Journal of Medicine and
            Philosophy 11: 385-404.
Kennedy, T.J. and Bennett, I.L.
    1981   The Planning of Research: The Role of the Federal Government. In Wechsler, H.,
            Lamont-Havers, R.W. and Cahill, G.F. (eds.) The Social Context of Medical Research.
            Cambridge, Mass.: Ballinger. Pp. 27-72
Kirby, M.D.
    1983   Informed Consent: What Does it Mean? Journal of Medical Ethics 9: 69-75.
Kopelman, L.
    1986   Consent and Randomised Clinical Trials: Are there Moral or Design Problems? Journal of
            Medicine and Philosophy 11: 317-346.
Lancet (Editorial)
    1984   Consent: How Informed? Lancet 1: 1445-1447.
Lellouch, J. and Schwartz, D.
    1971   L'essai thérapeutique: éthique individuelle ou éthique collective? Revue de l'Institut
            International de Statistique 39: 127-136.
Levine, R.J.
    1987   The Apparent Incompatibility Between Informed Consent and Placebo-Controlled Clinical
            Trials. Clinical Pharmacology and Therapeutics 42: 247-256.
Lhermitte, F. et al.
    1986   Double Blind Treatment of Chronic Multiple Sclerosis Using Hyperbaric Oxygen. Revue
            Neurologique 142: 201-206.
Macara, A.W.
    1986   Informed Consent. Lancet 2: 1101.
McEwen, J., Martini, C.J.M. and Wilkins, N.
    1983   Participation in Health. London: Croom Helm.
Marquis, D.
    1986   An Argument That All Prerandomised Clinical Trials are Unethical. Journal of Medicine
            and Philosophy 11: 367-383.
Matthews, W.B. et al.
    1985   McAlpine's Multiple Sclerosis. Edinburgh: Churchill Livingstone.
Medical Research Council.
    1948   Streptomycin Treatment of Pulmonary Tuberculosis. British Medical Journal 2: 769-782.
Meir, P.
    1979   Terminating a Trial: The Ethical Problem. Clinical Pharmacology and Therapeutics 25:
            633-640.
Moser, M.
    1986   Randomised Clinical Trials: Problems and Values. American Journal of Emergency
            Medicine 4: 173-178.
Neubauer, R.A., Schnell, R. and Goldberg, G.S.
    1979   Treatment of Multiple Sclerosis with Monoplace Hyperbaric Oxygenation. Paper presented
            to the 4th Conference on the Clinical Application of Hyperbaric Oxygen, Long Beach.
Neubauer, R.A.
    1980   Exposure of Multiple Sclerosis Patients to Hyperbaric Oxygen at 1.5-2 A.T.A. Journal of
            the Florida Medical Association 67: 498-504.

1983   Hyperbaric Oxygen Therapy of Multiple Sclerosis. Fort Lauderdale: Ocean Medical Center.

Nicholson, M. and McLaughlin, C.
1988   Social Constructionism and Medical Sociology: A Study of the Vascular Theory of Multiple Sclerosis. Sociology of Health and Illness 10: 234–261.

Prout, T.E.
1981   The Ethics of Informed Consent. Controlled Clinical Trials 1: 429-434.

Robinson, D. and Henry, S.
1977   Self-Help and Health. London: Martin Robertson

Robinson, I.
1987   Analysing the Structure of 23 Clinical Trials in Multiple Sclerosis. Neuroepidemiology 6: 46-76.
1988   Multiple Sclerosis. London: Tavistock.

Schaffner, K.F.
1986   Ethical Problems in Clinical Trials. Journal of Medicine and Philosophy 11: 297-316.

Silberberg, D.H.
1984   Problems in Evaluating New Treatments for Multiple Sclerosis. Annals of the New York Academy of Sciences 436: 418-422.

Simnett, J.D.
1982   We Need to Take a Fresh Look at Medical Research. Journal of Medical Ethics 8: 73-77.

Spodick, D.H.
1982   The Randomised Controlled Clinical Trial: Scientific and Ethical Bases. American Journal of Medicine 73: 420-425.

Teeling-Smith, G. (ed.)
1987   Measuring Health Status. London: Wiley.

Van Eys, J.
1982   Clinical Research and Clinical Care: Ethical Problems in the "War on Cancer." American Journal of Pediatric Hematology and Oncology 4: 419-423.

Vere, D.W.
1983   Problems in Controlled Trials: A Critical Response. Journal of Medical Ethics 9: 85-89.

Von Eickstedt, K.W.
1983   The Ethical Position of the Attending Physician in Phase 3 of Clinical Investigations. International Journal of Pharmacology, Therapeutics and Toxicology 21: 529-533.

Waldenstrom, J.
1983   The Ethics of Randomisation. Progress in Clinical and Biological Research 128: 243-249.

Webster, C. et al.
1988   Longterm Hyperbaric Oxygen Therapy for Multiple Sclerosis. Stansted, U.K.: Action for Research into Multiple Sclerosis.

Wiles, C.M. et al.
1986   Hyperbaric Oxygen in Multiple Sclerosis: A Double Blind Trial. British Medical Journal 292: 367-371.

Worthington, J.A. et al.
1987   A Double Blind Crossover Trial Investigating the Efficacy of Hyperbaric Oxygen in Patients with Multiple Sclerosis. In Clifford, Rose F. and Jones, R. (eds.) Multiple Sclerosis: Immunological, Diagnostic and Therapeutic Aspects. London: J. Libbey.

JOSEPH M. KAUFERT AND JOHN D. O'NEIL

BIOMEDICAL RITUALS AND INFORMED CONSENT:
NATIVE CANADIANS AND THE NEGOTIATION
OF CLINICAL TRUST

The signing of a consent agreement prior to surgery, invasive diagnostic or treatment procedures is a pivotal event in the negotiation of trust in the doctor-patient relationship. Most analysts have focused on the legal, ethical or procedural aspects of consent. However there is growing recognition of the need to consider political and cultural factors which lie outside the immediate context of the medical encounter and beyond the control of either physician or patient. This paper will examine the processes through which consent is negotiated when the patient is a Native from one of the remote areas of northern Canada.

In our research on cross-cultural health communication in urban hospitals, it was apparent that the negotiations that revolved around the signing of a consent form provided the clearest illustration of the unequal knowledge and power of the clinician and the patient. The clinician's approach to obtaining consent was based primarily on a biomedical understanding of a particular disease and associated treatment procedures; the approach of the Native patient to giving consent was based on experiential and cultural knowledge of past and present illnesses, interpretations of the social meaning of hospital regulations and health-professionals' behaviour, and general attitudes concerning intergroup relations in the wider society.

Legal approaches to informed consent and those rooted in biomedical ethics have not dealt adequately with situations in which clients and clinicians speak different languages and have different understandings of illness states and treatment options. This paper suggests an approach which is rooted in the social sciences and uses an ethnomedical and sociopolitical rather than an ethical or legal frame of analysis. A critical difference lies in the use of explanatory models which allow both the biomedical perspectives of the clinician and the lay or traditional perspectives of the client to be incorporated into the analysis. Research on the work of Native language interpreters provided unique access to situations of consent-giving in which differences in the culturally based explanatory models held by clinician and patient are critical to the process of negotiation. Yet, insight into the processes of obtaining consent requires understanding of the sociopolitical role played by a third party, the Native language interpreter. Seen within a social-interactionist perspective, language interpreters are cultural as well as linguistic intermediaries. Working in a situation involving consent agreements, the interpreter is a negotiator for the clinician who wants the trust of the client, and an advocate for the client who wants some understanding and control over what is planned by the clinician. Furthermore, in situations structured by historical relations of inequality, the medical interpreter emerges as an agent for empowering the patient to assert rights that might otherwise be ignored or denied.

41

G. Weisz (ed.), Social Science Perspectives on Medical Ethics, 41–63.
© 1990 *Kluwer Academic Publishers. Printed in the Netherlands.*

INFORMED CONSENT AS RITUAL:
ETHICAL AND LEGAL PERSPECTIVES

Medical ethicists, legal scholars and philosophers usually discuss informed consent in terms of idealized standards, using landmark legal decisions as examples and emphasizing the medical and legal mechanisms for effective management (White 1983). In common law, consent to medical treatment is a requirement with origins in rules governing surgical practice which held physicians liable for battery if they "touched" a patient without prior permission. The process of consenting to treatment emphasizes the principles of autonomy and self-determination protecting the client's right to exercise control over his/her own body (Faulder 1985). In governing physician behaviour, the doctrine emphasizes the principles of beneficence – or acting in the client's best interests – and non-maleficence – or duty to act with due care to avoid negligence and injury (Faulder 1985). For valid consent decisions to be achieved: (1) full information about risks, benefits and alternative treatments must be provided; (2) patient competency must be established; (3) patients must be able to understand the information presented; and (4) patients must be placed in situations in which they can act voluntarily (Meisel, Roth and Lidz 1977).

The historical evolution of ethical and legal criteria reflect the evolution of these principles. Prior to 1950, consent was primarily an agreement by the patient to proposed treatment in which the client freely conceded personal autonomy to the legitimate authority of the physician (Barber 1980). However since 1960, the concept of informed consent has been extended to apply to a wide range of treatment situations and formalized through a succession of North American court decisions (Fletcher 1983). A key U.S. decision (Nathanson v. Kline) established that physicians had the duty to reveal sufficient information about diagnoses, alternate treatment plans, and risks or benefits to enable patients to accept or reject proposed treatment. In treating "competent" adult patients, the physician was prohibited from imposing treatment, even if beneficent intervention was necessary to save the client's life.

During the past three decades the changes in the law of informed consent in the U.S. and Canada have modified the standards for determining whether a client is appropriately informed from more clinician-oriented to more consumer-oriented criteria which emphasize the need for "full disclosure" from the objective standard of the "reasonable patient" (Somerville 1981). Recent decisions have even evaluated the impact of the constraints on patients in situations where they were not provided with adequate information or were directly coerced (Lautt 1987). In both Canada and the U.S. the concept of the "reasonable patient," emphasizing what a competent client would need to know to make an informed decision, has emerged. The reasonable patient standard serves as a guide to physicians and courts in deciding which treatment alternatives and risks and/or benefits are salient to the patient and therefore must be disclosed (Hopp v. Lepp and Reibl v. Hughes).

Despite these trends, consent agreements continue to raise a wide range of ethical and legal questions which social scientists may address using alternative methodological and theoretical approaches. In much of legal and bioethics literature the actual decision-making process is ignored or described in formal, ritualistic terms. Although precedent cases are also framed within the context of adversarial relations, most do not convey a sense of the sequence of social interaction or a feeling for the problems of communication between participants who do not share the same sociocultural framework or socioeconomic status. Case studies therefore seldom provide information for assessing the power relationships between participants and the impact of wider contextual factors which influence consent proceedings. For these reasons much of the legal and clinical literature on consent does not adequately engage the issue of the development of trust relationships between patients and clinicians.

In emphasizing the formal or ritualized dimension of clinical, legal and ethical analysis of consent, we also recognize the positive functions of shared rituals. For anthropologists and sociologists, rituals are formal, repetitive behavior which express and renew basic cultural values and define social relationships (Helman 1984; Goffman 1967). Rituals portray in symbolic form norms governing interaction and establish the cognitive and linguistic categories through which members of a group or cultural system perceive social structure (Turner 1968). Much of the ethical, medical and legal literature emphasizes that formal consent agreements reinforce primary values governing profession-client relationships; these include self-determination, beneficence and trust (Faulder 1985). However, for consent rituals to reinforce common symbols and values defining healer-patient relationships, all participants must share a basic understanding of the structure and context of the interaction. Consent negotiations where clinicians and clients do not share common symbols or do not have common expectations about the nature of the doctor-patient relationship need to be examined in terms of their capacity to perform the integrative function of shared rituals.

The characterization of consent as a ritual act is not confined to jurists and ethicists. Clinicians and health consumers may also perceive the negotiation of informed consent as a formal act. Health consumer advocacy has concentrated on influencing written agreements by insisting that clinicians provide detailed information on diagnosis, treatment options and risks and/or benefits (Faulder 1985). Clinicians have become involved in providing more explicit information about areas of clinical uncertainty which were formally left implicit (Fox 1980).

More comprehensive requirements for information sharing are perceived by clinicians to have a dramatic impact on physician-patient relationships. Taylor and Kelner (1987) found that the majority of an international sample of oncologists regarded formal informed consent protocols specifying areas of risk and clinical uncertainty as an intrusion upon their relationships with their patients. Recent articles in medical journals respond to more client-oriented standards by focusing on clinicians' roles in "managing" the negotiation process and proposing methods for satisfying formal professional and legal criteria necessary to avoid litigation (Gray 1978; Meisel and Roth 1980).

Despite the emphasis on co-participation and autonomy in both consumer and medical-legal literatures, clinicians retain very real power in the consent negotiation. Their power is based partly on their exclusive access to biomedical knowledge, partly on their hierarchical position within the health care system, and partly on their ability to withhold treatment. In decisions involving physicians and patients sharing Euro-Canadian cultural background, both participants may be able to refer to a common body of lay knowledge defining illness, treatment, and the organization of medical care. When consent agreements are negotiated in cross-cultural situations, the absence of shared concepts of illness or treatment as well as the absence of common values regulating healer-client interaction amplify the power imbalance.

Unfortunately the more formal examinations of sources of power in the legal, ethical and clinical literature tend not to consider how differential power is actually expressed in ritual interaction sequences. Legal and ethical analyses of consent decisions often decontextualize the process of negotiation from the cultural, organizational and sociopolitical environment of the health care system. In this literature, consent agreements are often described as though the client-clinician relationship could be removed from the context of hospital structure and socioeconomic barriers to participation. In keeping with the editorial focus of this volume, we will utilize social science perspectives to extend the analysis of consent decisions in order to address these wider issues.

ETHNOMEDICAL AND SOCIOPOLITICAL PERSPECTIVES ON INFORMED CONSENT

Relatively little social science research has actually examined the process of interaction in consent negotiation. We will, therefore, initially examine the alternative insights offered by two complementary analytic frameworks developed by medical anthropologists and medical sociologists.

Contemporary medical anthropologists have developed ethnomedical frameworks for interpreting illness behaviour and health related decisions (Kleinman 1975, 1980; Good and Good 1981). Analysis of informed consent decisions involving clients and clinicians from different linguistic and cultural backgrounds provides an opportunity to examine systematically the impact of cultural reference frameworks. Negotiation of consent in cross-cultural encounters may also provide a model for understanding the more general processes of interpretation and mediation which occur in all clinician-patient interactions. Among members of the same cultural and linguistic group, differences between clinicians' and patients' understandings of illness and proposed treatment may reflect the existence of different lexicons, the use of similar terms embedded in different nosologies, or the use of the same term to refer to alternative disease entities and illness concepts (Harwood 1981). Consent decisions in all clinician-patient encounters involve processes of establishing a shared framework for understanding. However, the process of mediating between client and professional explanations of illness, and proposals for treatment, becomes more complex when the participants come from different cultural and linguistic backgrounds (Matthews 1983).

Medical anthropologists have developed the concept of explanatory models as a theoretical and methodological device to document culturally based assumptions about illness and treatment. Kleinman (1980) has defined explanatory models as culturally based "notions about an episode of sickness and its treatment that are employed by those engaged in clinical processes, whether patient or clinician." Individual and shared explanatory models shape clients' and clinicians' understanding of disease etiology, symptomatology, pathophysiology, natural history and guide proposals for treatment (Helman 1985). Contemporary analysis of clinician-patient interaction stresses the use of explanatory models to document and reconcile the ways that all participants perceive disease entities and treatment options (Katon and Kleinman 1981; Hahn and Gaines 1984).

In idealized models of informed decision-making, the clinician is expected to elaborate treatment options, explain the patient's right to refuse treatment and offer information about the risks and benefits of alternative treatments. In consent decisions, clinicians frequently follow up diagnostic explanations with a description of other diagnostic and treatment options using biomedical terminology. To provide the basis for meaningful choice, the clinician or an intermediary must first translate biomedical terms and concepts into more accessible explanations. Initial translation of biomedical concepts into lay language may never take place, or may occur when questions from the client force the physician to expand and clarify their initial explanation (Tuckett and Williams 1984). In situations where interpreters or other intermediaries are present, translation of biomedical terminology into lay concepts may occur as part of the process of language and cultural interpretation (Kaufert and Koolage 1984). For optimal clinician-patient communication, explanations of treatment options should involve mediation between clients' and clinicians' explanatory models of risk and benefit (Katon and Kleinman 1981).

Our observation of the negotiation of consent agreements involving Native patients revealed that clients were frequently presented with a limited subset of treatment or diagnostic options which the clinician priorized on the basis of his or her assessment of the best clinical option. Actual cases of true co-participatory decision-making in which all options were discussed and relative risks and benefits presented were seldom observed. These observations are consistent with the wider critique of the explanatory model approach. Despite the strong appeal of ethnomedical approaches, explanatory model frameworks have been criticized for their limited capacity to relate client and clinician beliefs to actual interaction patterns. Ethnomedical frameworks have also been criticized recently from a critical theory perspective emphasizing either political economy (Young 1982; Taussig 1980) or deconstructionist approaches (Scheper-Hughes and Lock 1987). These critiques suggest that people's interpretations of events and the meanings they assign to those events are structured by their historical participation in the social hierarchy and particularly by their access, or more importantly, lack of access to the construction of ideologies. In other words, individuals do not freely choose their beliefs and values from a cultural smorgasbord, but rather their "explanatory models" are a product of historical and ideological relations of economic and social inequality.

Interactionist frameworks developed by medical sociologists and extended by medical anthropologists may provide insights for examining the dynamics of consent negotiation in a sociopolitical environment. Symbolic interactionist frameworks for client-patient encounters focus on what is happening in the encounter and how actors influence each other's perception of the situation (Lazarus 1988). Interactionists have examined the relative power and potential for social control in terms of differential access to information (Stimson and Webb 1975) and the influence of external political structures (Waitzkin and Stoeckle 1976). This asymmetry in clinician-patient relationships has also been explained in terms of sources of professional dominance and control over technology (Freidson 1970; Zola 1981). Inequalities in access to each of these sources of power mean that the clinician is most often more autonomous and the client more dependent. Although ethnomedical approaches to physician-patient interaction stress the need to redress power imbalances (Katon and Kleinman 1981), co-participatory decision-making may be inherently alien to cultural assumptions of biomedicine (Lazarus 1988). Egalitarian ideals in healer-client relationships may also not fit the expectations of certain patients.

Neither interactionist nor ethnomedical analysis of clinician-patient encounters has systematically examined the impact of intermediaries, such as interpreters, family members or advocates upon clinical communication. From our research on Native medical interpreter-advocates, it is clear that they play a profound role in mediating the relationship between clinician and clients (Kaufert, O'Neil and Koolage 1986). In negotiating patient agreement to treatment or diagnostic options, language translators and family members frequently add another level of interpretation in which the intermediary introduces his/her own explanatory models or influences the level of information available to client and physician. Often, additional information provided by the interpreter incorporates the patient's previous experiences in the home community or urban migrant household and in hospitals. Native patients' decisions then, were more often than not based on limited biomedical information combined with a broader personal ethnomedical and sociological understanding of the illness event. Unfortunately, few clinicians were aware of the cultural and sociopolitical construction of the consent decision, since it usually was negotiated in a Native language.

The experience of Native Canadians in urban hospitals provides an opportunity to examine consent decisions in cross-cultural interaction. Consent agreements bring into sharp relief the differences between biomedical and Native interpretations of cultural content and power relationships in clinician-client interaction. They also provide a context for documenting the impact of intermediaries upon clinical communication.

METHODS

The data base and methodological design for analysis of consent decisions were derived from a long term research program on medical interpretation in urban hospitals and health communication in Inuit communities of the Canadian Arctic (O'Neil 1988; Kaufert and Koolage 1984). The negotiation of informed consent in interactions between Native clients and clinicians was documented through recording case encounters in medical wards and outpatient clinics (Kaufert et al. 1984). Because of our focus on medical interpreters, the majority of observations were limited to consent decisions in which interpreters were used. However, some encounters were recorded in which family members, rather than trained interpreters, provided translation.

Field observations documented the work of eight Cree, Ojibway and Island Lake language-speaking interpreters working in two Winnipeg teaching hospitals. About 25 cases involving consent agreements for surgery or invasive diagnostic tests were observed by the investigators and two research assistants with fluency in Native languages. Field observations were followed up with focused interviews with patients and attending medical or nursing staff. Half the consent proceedings were videotaped and translated into verbatim English text to facilitate more detailed analysis of the impact of language usage and sequence of interactions. Follow-up interviews allowed us to document the varied perceptions of the consent agreement from the standpoint of the interpreter, the client and the clinician. The impact of the hospital organization and professional ideology on clinicians and medical administrators was documented through observing clinical staff meetings, case conferences, and medical rounds.

The remaining sections of this paper will be devoted to an analysis of two case studies of consent decisions involving Native patients, health professionals and medical interpreters. Both explore the potential contribution of ethnomedical approaches to understanding the problems of reconciling client and practitioner interpretations of illness and treatment alternatives. Each one also examines the process of social interaction in consent negotiation within the broader sociopolitical context of power relationships.

Case Example I: Mediating Client and Physician
Explanations of Invasive Diagnostic Procedures

The first case involves negotiation of an informed consent agreement for a Native patient to undergo gastroscopic and colonoscopic examinations. The 46-year-old Cree-speaking woman, was referred from a northern nursing station for further investigation of anemia by a gastroenterologist in Winnipeg. Encounters centering on consent negotiation and diagnosis were videotaped at each stage of the diagnostic workup. In each encounter the physician worked with a Cree-speaking medical interpreter to explain diagnostic and treatment options and negotiate patient consent for examinations of the stomach, small and large intestine.

In the initial encounter, the physician attempted to evaluate the patient's understanding of her own problem and to explain his diagnostic model of the probable cause of anemia. Specifically, he attempted to move from discussing the client's understanding of anemia (conceptualized by the patient in terms of weakness) to a more complex model linking the loss of blood to the presence of lesions caused by anti-inflammatory medication. Following a cursory explanation of the general diagnosis, the physician moved to a series of diagnostic questions about presence of blood in the patient's stool.

Doctor: She's anemic and pale, which means she must be losing blood.
Interpreter: (Cree) This is what he says about you. You are pale, you have no blood. [Cree term for anemia connotes bloodless state.]
Doctor: Has she had any bleeding from the bowel when she's had a bowel movement?
Interpreter: (Cree) When you have a bowel movement, do you notice any blood?
Patient: (Cree) I'm not sure.
Interpreter: (Cree) Is your stool ever black, or very light? What does it look like?
Patient: (Cree) Sometimes dark.

At this point the patient told the interpreter that she did not understand how her "weakness" (anemia) was related to questions about gastro-intestinal symptomatology in the physician's reference to dark stools. Without asking for additional explanation from the physician, the interpreter attempted to link the patient's understanding of her anemia with the concept of blood loss.

Interpreter: (Cree) We want to know, he says, why it is that you are lacking blood. That's why he asked you what your stool looks like. Sometimes you lose your blood from there, when your stool is black.

In discussing the probable etiology of the woman's anemia, the physician introduced a complex explanatory model which explained gastric or intestinal bleeding in terms of the possible side effects of anti-inflammatory medication for rheumatism. The patient again indicated that she did not understand why the questions about her experience with medication for rheumatism were relevant to the current diagnosis of problems of weakness and blood loss. The interpreter provided an unprompted explanation linking the line of questioning about the side effects of anti-inflammatory drugs with the concept of blood loss.

Interpreter: (Cree) He says that those pills you are taking for rheumatism, sometimes they cause you to bleed inside, or you will spit up blood. Not everyone has these effects. This is why he wants to know about your medication.

Following gastroscopy, the physician attempted to explain the results of the gastric studies and at the same time to extend the initial consent agreement to permit colonoscopic examination of the lower bowel.

Doctor: Everything looked good. There was no ulcer and no nasty disease in the stomach or duodenum or the esophagus. No bleeding. You're still anemic so we still want to find out if there's any bleeding from the lower end.
Interpreter: (Cree) He says this about you: there's nothing visible in your stomach. Nothing, no sores, lumps, what they call "ulcers." Nothing from where you swallow. Nothing wrong that can be seen.
Patient: (Nods, but makes no verbal response).
Doctor: We're going to put a small tube in from the colon. It's only this big. To have a look, to see if there's any abnormality. It won't take too long and it will be very quick and you shouldn't be uncomfortable with it at all. Okay?... So we'll go ahead and do that now while we can.
Interpreter: (Cree) He wants to see you over here from where your bowel movements come from. Something will be put there, like the first one [the tube you swallowed], but smaller. So you can be examined "down there." Maybe somewhere "down there," it'll be seen that you are losing blood from there. The reason why you are lacking blood. That's what he's looking for. The cause for your blood loss.

During the exchange the patient's willingness to extend the initial signed consent document to cover the investigation of the lower bowel was inferred from a nod and no real alternative was discussed by the physician. The patient was asked to initial the addition to the consent agreement, without formal translation of the English text.

Colonoscopic and radiological examination of the intestine revealed a benign polyp. The physician felt that the polyp should be cauterized in a second colonoscopy and asked for the patient's consent for the additional procedure. Although risks and benefits were not formally discussed, the interpreter elaborated on the basic diagnostic information provided by the physician. The interpreter also introduced a more formal decision point at which the patient was asked to give her formal consent.

Doctor: We x-rayed the bowel.
Interpreter: (Cree) And this is what they did this morning when you were x-rayed. The pictures of the area you have bowel movements.
Patient: (Cree) Yes.
Doctor: And that shows a polyp, a small benign tumor. And I have to take that out.
Interpreter: (Cree) This picture they took this morning. He saw it already. There's something growing there. About this size. And it has to be removed, because you might bleed from there.
Patient: (Cree) Yes.
Doctor: Now I can take it out without an operation, by putting a tube inside the bowel, and putting a wire around it and burning the polyp off. That stops the bleeding, no need for an operation.
Interpreter: (Cree) He says they can put in a tube like before and burn off the growth.
Patient: (Nods but makes no verbal response).

Doctor: If she wants to make the arrangements for the hospital admission she can come down and sign the consent form.
Interpreter: (English) Will there be complications?
Doctor: There are a few complications but I think it would be difficult to explain them all.
Interpreter: (Cree) After this procedure has been done you won't be staying here at the hospital. You'll be able to go home on Saturday. It will be done on Friday, then you'll already be able to go home on Saturday. It won't be long. But it's entirely up to you.

In this exchange, the interpreter is providing more than a simple translation elaborating on the risks, benefits and rationale for the procedure. The physician assumes that his statement about the necessity for the procedure will be sufficient to obtain consent. He further assumes that it would be too difficult to explain all the possible complications. However, in both instances, the interpreter does not provide a literal translation of the physician's side comments, but attempts to assure the patient and justify the procedure through explanations incorporating concern with the length of her stay and desire to be with her family. The interpreter assures the patient, "you'll be able to go home on Saturday" emphasizing the expectation that the operation will be minor and that she won't be separated from family and home for long. The interpreter's statement linking approval of the consent to the patient's early return to her community occurs in Cree and therefore is not accessible to the physician.

At the end of the encounter the physician included the consent agreement with the hospital admission protocol and assumed it would be signed with the other paper work. The interpreter provided a more direct opportunity for the patient to give or withhold consent.

Interpreter: (Cree) Do you want to have this procedure done? Will you consent to have this growth removed burned? Do you consent to have it done?
Patient: (Cree) I don't know.
Interpreter: You know, if it's not removed it may bleed. It may cause problems.
Interpreter: (English directed at physician) Dr. ___ , isn't it true that if it's not removed, it can bleed and she can become anemic?
Doctor: That's correct, we feel that your anemia may result from the bleeding of the polyp.
Interpreter: (Cree) If it's not removed, you may end up with cancer. You know? And you will not have an operation. It's harder when a person has an operation. You know? And [this procedure] that he's going to do will get it on time. Before it begins to bleed or starts to grow. You're lucky it's caught on time. And it will bother you when you have a bowel movement. This way there's no danger that this growth will bleed.
Patient: (Cree) I still don't know.
Interpreter: (Cree) Well if you want to come in for the procedure while you are here? It's all up to you to think about.

Again, the interpreter has assumed responsibility for providing a rationale for the procedure and explaining the potential benefits. Her explanations are also clearly based on her own medical knowledge, and her understanding of the patient's explanatory model. The fear of cancer in the Native community is linked to a general understanding of the history of infectious disease epidemics that nearly destroyed aboriginal society in North America. Cancer is increasingly viewed as the new "epidemic." The interpreter is using her knowledge of these fears to negotiate the patient's consent, but she is also using Cree models of negotiation emphasizing individual autonomy. Her final statement emphasizes her client's ultimate personal responsibility: "It's all up to you to think about."

At this point, the patient accompanied the interpreter and physician to the appointment desk and scheduled the colonoscopy for the following day. After a brief summary of the text (which was printed in English) was provided by the interpreter, the patient signed the consent form. The formal act of signing the form was immediately subordinated to a discussion of specific arrangements for the client's discharge from hospital and travel arrangements for returning to the reserve community.

In this case, consent was negotiated by drawing on the patient's trust in her relationship with the interpreter. The physician initially assumed that little explanation was required for consent, and indicated his unwillingness to negotiate. Information sharing occurred gradually over the course of the encounter. The interpreter assumed the negotiator's role, based on shared cultural understanding of both biomedicine and Native culture. Throughout the sequence of interaction, the interpreter worked to elicit and clarify both the client's and clinician's interpretation of the condition and the program of treatment. However the interpreter's intervention introduced a third party into the clinician-client relationship. She directly influenced the course of the decision by independently introducing new information about illness and treatment options. She also imposed decision points where the patient could actually exercise her option to consent.

The patient's willingness to allow the interpreter to negotiate consent is also evident in her reluctance to ask the physician direct questions. In response to the interpreter's questions about her understanding, she repeats at several junctures, "I don't know." This provides a cue for the interpreter to introduce further information or elaborate relative risks. The final consent is passive, in the sense that the patient signs the forms without further resistance.

Case 2: Strategic Interaction in Interpretation of Consent for Surgery

A second example of consent negotiation was observed in a situation in which interpreters were asked to translate for a Native patient after she had signed a written agreement for surgery. In this case, the role of the interpreter in providing the patient with opportunities for informed choice and control was limited by (1) the situational context of an evolving medical emergency; (2) the intervention of family members as proxies and informal translators; (3) the clinician's maintenance of strategic control over the interaction; and (4) the definition of the client's physical and cognitive "competence" to make informed decisions.

This case involved a 74-year-old Cree-speaking woman who had been transported from a remote reserve community for vascular surgery in an urban hospital. The client had been unsuccessfully treated in a regional hospital for an "ischemic left foot," and had had two toes surgically amputated. The woman was accompanied by family members and represented by her daughter who had resided in the city for several years and had some experience as a translator. The client had minimal command of English and initially relied upon her daughter to translate for her during the intake examination and pre-surgical consultation. Language interpreters from the Native Services Department of the hospital were not initially asked to translate for the patient until the time when vascular surgery was about to be performed. The proposed surgery involved a vascular graft bypassing the ischemic area to restore circulation to the lower extremity.

As the medical interpreter began to translate for the woman in pre-surgical consultation, she recognized that the client had little understanding of the proposed vascular surgery and did not relate the intervention to her long term circulatory problems or current infection. When the interpreter asked the Head Nurse whether she needed to help translate the consent form, she was told that it had already been signed during a consultation between the surgeon and the daughter earlier in the day. The patient further indicated to the interpreter that her daughter and the physician had talked for an extended period in the hall, but had not discussed the details of the surgery with her. The patient stated: "My daughter didn't tell me anything."

The interpreter attempted to provide a more adequate explanation of the procedure, its risks and the type of anesthesia which would be used. She began by calling the daughter who had been involved in translating the consent agreement. The interpreter explained that she felt the patient did not sufficiently understand the surgical procedure which was going to be performed. The daughter indicated that she had tried to simplify the explanation of the vein bypass given by the doctor using appropriate Cree terms. The daughter also emphasized that she had provided a very general explanation because the physician felt that she should avoid frightening her mother.

Following the telephone conversation with the daughter, the interpreter called the surgeon. The physician emphasized that the daughter had acted as a proxy in the consent, although the woman had signed the agreement herself. The physician felt that the primary objective was to communicate with the daughter because the patient had hearing impairment and minimal command of English. The doctor concluded that sensory impairment, problems of language interpretation and deteriorating medical condition compromised the patient's capacity to make informed decisions and had therefore elected to involve the family.

The following day the interpreter was visiting with the patient and the surgeon when the anesthesiologist entered the patient's room for the final pre-operative consultation. The interpreter asked the physicians to briefly summarize the diagnosis and proposal for treatment.

Interpreter: (to the physician) Could you tell me again what you are going to do; because I wasn't here when you explained it before?
Doctor: We are going to do a bypass.
Interpreter: What kind of bypass? Is it similar to heart bypass?
Doctor: No, we're going to take some vein from here (pointing to the thigh) and put it in here (pointing to the lower leg).
Doctor: We need to do vascular surgery because of the recurrent infections which made it necessary for them to amputate her toes. There is no way this lady is going to heal without a bypass.
Interpreter: Could you explain some of the details?
Doctor: Just emphasize that we're going to take a piece out here and put it there and then it will heal. I don't want to go into the details she [the patient] knows about it. The daughter was here interpreting last night and knows all about it. They already signed the consent form.
Interpreter: (Cree) This foot of yours, where your two toes were removed, you can't seem to heal because there's nothing getting through. You have no feeling there because your blood doesn't reach there. You can't heal because you have no feeling. Like, when someone is numb, something is blocking the way, maybe a vessel [is blocked], but it is not for certain what it is. They are going to take part of your vessel from your thigh here and it's going to be sewn into here by your foot where it's blocked. It will be cut there then the other will be placed there. Then your blood will be able to flow properly. Then you'll heal.

The physician left the patient's room, leaving the interpreter to follow up her explanation of the surgery in terms of its effect on vascular insufficiency and potential for avoiding recurrent infections. This explanation again was provided outside the context of the physician-patient encounter. As such it highlighted the interpreter's role as the "curator" of the patient's personal history. By accompanying the patient in encounters with several clinicians and interacting in informal situations, the interpreter established a basis for information sharing and trust outside the context of the medical encounter.

The vascular surgery was performed the following day. When the interpreter visited the patient on the post-operative ward, she found the woman was profoundly depressed. The patient indicated that neither the physicians nor her daughter had prepared her for the amount of post-operative trauma or for the magnitude of the surgical wound (extending from her upper thigh to her lower ankle). Two days following the surgery the patient's condition deteriorated as the leg became gangrenous. The emergency necessitated an immediate consultation to obtain consent for surgical amputation of the limb. The surgeon recontacted the family, but indicated that the patient was too "ill and confused" to make the decision without assistance from other family members. The surgeon also explained that the initial consent agreement for the bypass surgery included a section advising the patient of the potential risk of amputation and authorizing the procedure in emergency situations.

The interpreter was again involved in translating for the physicians, patient and family members in the pre-operative consultation. The interpreter found that the patient was conscious and able to engage in a limited, but rational conversation. The surgeon's initial comments emphasized that the risk of surgical amputation had been discussed in the consent protocol for vascular surgery. The interpreter indicated to the physician and the daughter that she felt the woman would not consent to amputation even if this diminished her chances of survival. The interpreter indicated that the patient understood that amputation was the only hope of saving her life, but was reluctant to consent because of her traditional beliefs about maintaining her corporal integrity after death. Cree beliefs stress the need to maintain all parts of the body intact prior to and for several days after death.

The surgeon listened to the interpreter's elaboration of the woman's explanatory model. However, he insisted that consent for amputation was already part of the initial approval. He emphasized the patient's diminished capacity to make an informed decision. The surgery was performed but the patient died three days later without regaining consciousness.

This case study clearly illustrates the relative influence of ethnomedical and sociopolitical factors in shaping clinical decision-making. The professional interpreter's partial knowledge of the patient's explanatory model did not derive from direct doctor-patient communication, but rather combined information from private conversations with the patient with information based on her personal knowledge of the cultural context. In this case, the interpreter was doing more than sharing understandings with the patient that even the patient's family were unaware of. The interpreter may also have been projecting some of her own beliefs and values in articulating a collaboratively constructed explanatory model emphasizing the traditional notion of maintaining corporal integrity.

The interpreter was able to introduce only selected ethnomedical information into a dialogue controlled by a physician because the latter assumed that there was only a single possible course of appropriate clinical management. The physician had two options when confronted with ethnomedical information which was contrary to his broad interpretation of the medical conditions of the case. He could return to the patient and, through the interpreter, ask the patient to elaborate on this new information. This type of intervention would have raised the possibility of reassessing the medical decisions. The physician could also treat the ethnomedical information as extraneous to the medical facts which guide and control his ultimate clinical decision. This second option was apparently taken by the physician in Case 2.

Another factor limiting the capacity of the patient to make informed decisions was the rapidly evolving sequence of life-threatening problems which characterize medical emergencies. The definition of a treatment situation as an emergency establishes a situational context which justifies delegation of decision-making, limitation of treatment options, and imposition of time constraints. In the case example, both consent for the initial bypass surgery and final amputation occurred in circumstances defined as "emergencies." The immediate life-threatening character of the patient's problem reinforced the surgeon's decision to make the decision himself, using a family member as a proxy. In a follow-up interview the surgeon gave his rationale for presenting a limited inventory of treatment options and not using a professional interpreter. The physician stated:

There is a language gap but, in a situation like this, I have no problems. It is not as in the case of cosmetic surgery because it is a case of limb salvaging. There is no alternative. If the surgery is not done, they are going to lose a leg. So, I don't have any problem convincing them because I don't operate on somebody where this is not necessary.

Emergency situations were interpreted by the clinician as limiting the range of available clinical options which he felt obliged to discuss with the patient.

The involvement of the patient's family in both language interpretation and in final decision-making illustrates the extent to which designation of proxy decision-makers may be influenced by linguistic and cultural access rather than diminished cognitive capacity for informed decision-making on the part of the client. The designation of family members as proxies for unconscious, senile or young children occurs routinely, although it has no clear status within Canadian law. Again, issues of cultural and linguistic access complicate the issue of the use of family members as proxies. The use of untrained translators from the patient's family group raises several other issues including: the problems of assuring objectivity in translation; the problems of substituting proxy decision-makers for psychologically competent patients who do not speak English; and the problems of assuring that patients are given access to professional interpreters early in the sequence of care decisions.

In the case of this patient undergoing vascular surgery, the woman's experience illustrates the impact of using volunteer interpreters. The surgeon subsequently stated that he preferred to involve family members as translators for elderly Native clients because this procedure minimized their opportunity to complain about treatment outcome. His use of a family member as a translator was also not regulated by the ethical guidelines or professional standards governing interactions between professional interpreters and their clients. The family members may be influenced by personal emotions and the agendas of their kin group. Their role in communication with the patient is seldom evaluated in terms of either linguistic adequacy or impartiality.

In interpreting for the surgeon and the patient, the daughter incompletely translated the biomedical explanation of the bypass surgery and amputation. She also failed to elicit from the patient and to communicate to the physician the woman's own traditionally based belief that the loss of a limb would compromise her ability to function in the afterlife. The daughter's willingness to assume a proxy role complemented the surgeon's control of the sequence of decisions.

This case also highlights the tendency to utilize linguistic or cultural differences to justify labelling patients as having "limited capacity" to make consent decisions. Our research suggests that prior to the establishment of professional interpreter programs in the hospital, consent decisions for Native patients who did not speak English were made routinely as if the patient had "limited cognitive capacity." Indeed, this case clearly shows that despite the availability of professional interpreters, many medical practitioners still regard Native patients as if they were "children, senile or unconscious."

POWER AND CONTROL IN CONSENT DECISIONS AMONG NATIVE CANADIANS

These two case studies illustrate that in Native-Canadian patients' encounters with the urban health care system many of the key tenets of consent doctrine are subject to cultural and class bias. The application of the "reasonable patient" standard assumes that the client has linguistic access and some common framework for cultural interpretation of diagnostic and treatment information. Legal standards do not fully control for barriers to informed clinical communication arising from the cultural background and language orientation of client and clinician. Criteria for establishing diminished capacity have been liberalized to strengthen the rights of pediatric and psychiatric patients. However, as the second case illustrates, lack of linguistic or cultural access among Native patients often leads to the delegation of treatment decisions to a proxy, such as a family member, community elder or member of the attending clinical staff.

Both cases clearly show that culturally grounded premises behind the consent agreement create the potential for cultural conflict. Medical and legal models for informed decision-making assume that the clinician and client share common assumptions about the illness, the nature of the healer-patient relationship and significance of risks and benefits attached to alternative courses of action (Lautt 1987). However, Native patients may reject biomedical explanations of the pathophysiology of an illness in favour of more holistic approaches stressing spiritual causation (Kaufert and O'Neil 1988). For example, we documented another case in which an Ojibway woman rejected a pediatrician's diagnosis of her child's illness as a genetically transmitted syndrome. The mother interpreted the etiology of the child's illness in terms of her own transgression of customary law and insisted that only her participation in traditional healing rituals could cure the child's illness.

Native cultural values may also conflict with medical legal assumptions about power relationships and information exchange in the clinician-patient relationship (Jakobovits 1983). Current emphasis in legal and consumer literature on co-participation and egalitarianism in consent decisions may even conflict directly with Native cultural expectations about healer-client relationships. Relationships between Native patients and traditional healers do not emphasize two-way information exchange leading to mutual agreement about the etiology of a condition and its treatment. Rather Cree and Ojibway patients were expected, at the time of their initial contact with the healer, to commit themselves to follow his or her instructions without question. This difference in expectations was observed in consent decisions involving several Native patients who indicated that the physician should simply state what was wrong with them and prescribe a single treatment regimen, rather than discussing all of the alternatives.

Because of differences in their cultural models of healer-patient relationships, Native clients may appear to practitioners as being passively involved in their medical care. This presumption reinforces the physician's perception that consent will be given automatically without negotiation. However, our data suggests that "passivity" or a willingness to submit to the authority of the physician, has a very different meaning for Native patients. Despite the efforts of medical interpreters to broaden the Native patients' participation in the consent process (as described in the case studies) many consent agreements were signed without question, despite concerns that the patient may have had about the impact of such interventions on their personal lives. Analysis of interactions described in both case studies suggests that the cultural basis for apparently passive acceptance of clinical trust relationships is often a complex mix of traditional understandings about appropriate roles and behaviours for patients and healers, and attitudes of resignation and alienation arising from the historical experience of a colonial medical system.

Our Native informants indicated that Native healers expect absolute compliance from their Native patients and often will not accept patients who do not bring this submissive attitude to the traditional healing encounter. As one Elder stated:

The Indian doctor expects his patients to have already decided they will do whatever they are told when they come to see him. By offering tobacco, the patient is asking the Indian doctor to help him and he is indicating he will *obey* all instructions.

Indian culture does not allow for the possibility that a healer would deceive or lie to a patient, since the healer's decisions and actions are guided by spirit helpers.

Within Cree and Ojibway culture, when people contact a healer they do so with an attitude of humility. They are asking the healer to assist them in restoring balance to their lives and they must be prepared to make sacrifices in order to achieve this. The healer's first task is to assess the sincerity of the patient's request and the degree of preparedness towards accepting the healer's prescriptions without question. If a healer considers a patient to lack commitment or to be unprepared, s/he can and does refuse to provide treatment until the appropriate attitude is expressed by the patient.

In Indian culture, the contract between the healer and the patient is symbolized by the offering and acceptance of tobacco. Tobacco is a sacred substance provided to Indian people by the Creator to facilitate communication between humans and the spiritual world. By offering tobacco to a healer, a person is demonstrating his or her respect for the healer's authority and willingness to accept the healers advice. The tobacco offering then, is the culturally equivalent ritual to the signing of a consent form in "White man's" medicine.

The Native patient's willingness to accept the authority of the Western physician is not, however, derived solely from cultural expectations regarding a healer's authority. The history of Western medicine in Canadian Native communities is one of colonialism; and the concept of informed consent in medical treatment for Native patients is very recent (O'Neil 1988; Culhane Speck 1987). This historical relationship was dramatically illustrated in the late 1970s when a number of Native women complained that they had been sterilized without their knowledge. In the ensuing public debate, replete with accusations of genocide, it emerged that tubal ligations were being performed on older multiparous women when the obstetrician determined that it was in the patient's medical interest. Although consent agreements were available, most women indicated that they had simply agreed to whatever the nurse or doctor "told them to do," because they had always done so in the past. The politicization of the issue occurred when the church and Native leadership re-interpreted a medical event in the context of colonial relationships (Lechat 1976/77; Baskett 1978).

Although Native patients' willingness to accept a passive role in contemporary medical consent rituals is probably rooted in traditional values, the recent history of medical experience with Western physicians and nurses has provided little opportunity for a more informed and participatory relationship to emerge. In this context, the role of the Native medical interpreter-advocate to empower Native patients in consent negotiation is crucial.

Another dimension of consent decisions illustrated by the case studies involves communication of information about material risks and benefits of alternative treatments. For Native Canadian and Euro-Canadian patients alike, the concepts of risk and benefit are fraught with ambiguity (Lautt 1987). Definitions of relative risk and benefit interpreted in terms of statistical probabilities may have little personal meaning in the process of deciding between alternative treatment modalities. In case 1, the physician refused to provide detailed information about the potential complications of diagnostic procedures and surgical interventions. In the second case, risk vs. benefit information was not presented to the patient, the interpreter or the family because the physician believed that an emergency situation and absence of treatment options limited him to a single course of action. In another case documented in our health communication study, the parents of a Native pediatric patient with metastatic cancer were informed that they could improve the child's survival chances by consenting to limb amputation. The biomedical interpretation of survival probabilities could not be reconciled with the parents' traditional belief that amputation would profoundly compromise the child's ability to function without a leg in an afterlife.

For Native clients, cultural barriers to informed decision-making are compounded by barriers arising from the socioeconomic and organizational structure of the Canadian health care system. For half of the 120,000 Manitobans who claim Native ancestry, health care in Northern and remote communities is provided by a complex system of federally funded health centres (nursing stations) and regional or urban hospitals. The provincially supported tertiary care system involves most Native patients in a complex network of referral and transportation to urban centres where they are likely to be isolated from the support of family and community.

Hospital-based systems of diagnosis and triage, fragment and compartmentalize the patient's relationship with treatment staff. It may, therefore, be more difficult to establish personalized relationships which are considered by Native patients to be necessary for the establishment of interpersonal trust. Trust relationships in turn were felt by many Native people to be the primary feature of their consent to treatment. Native patients in urban hospitals have less power and autonomy in clinical interactions because of the lack of cultural and linguistic capacity for interpreting and manipulating organizational roles and clinical relationships (Kaufert and O'Neil 1988).

Feelings of powerlessness in consent negotiations for Native people may be heightened in hospital settings by lack of patient understanding of the hierarchical relations among medical staff and emphasis upon unfamiliar technology. As case 2 illustrates, initial consent agreements may also be interpreted as open-ended commitments to comply with subsequent clinical decisions. Most Native patients sign initial consent agreements in the nursing station before referral for tertiary treatment or diagnosis in urban hospitals. This initial consent agreement may be sequentially expanded in the tertiary care hospital as the diagnostic and treatment interventions are elaborated. Lautt describes this process:

> As the diagnosis of the hospital unfolds, the level of complexity of his condition increases, while the sources of information with whom he is familiar and can trust decrease, encouraging passivity and a fatalistic acceptance of the processes around him (Lautt 1987: 85).

The expansion of the terms of consent may occur as the patient is referred into specialized hospital wards where she or he has minimal information about and control over treatment. In some cases administrative and economic pressure may even be brought to bear by administrators who may refuse to support transportation and living expenses unless the patient consents to the proposed treatment. The combined impact of sociocultural barriers with these structural features of the health care delivery system raises the alternate question of whether any consent agreement involving Native Canadians can be informed in the sense of reflecting shared understanding, co-participation and clinical trust.

The case studies clearly demonstrate the role of interpreter-advocates in redressing cultural and structural bias against Native people in consent negotiations in urban hospitals (Kaufert and Koolage 1984). In the two hospitals in which the case encounters were observed, interpreters have expanded their role beyond language translation to advocacy roles.

Both case studies indicate that the role of interpreter-advocates in the consent negotiation cannot be separated from the wider context of their work with the patient and community. In both case studies interpreters were constant sources of information about illness, diagnostic and treatment options, hospital procedures, and patient rights. They provided support to the patient, the family and home community. The presence of the language interpreter or patient advocate as an intermediary in the negotiation of consent raises a number of ethical and procedural issues. The involvement of interpreter-advocates in the doctor-patient relationship may shift responsibility and initiative for disclosure from the client and clinician to the intermediary. As the first case study illustrates, the translator may exercise control through selective interpretation of information provided by the client. The interpreter may also play an alternative role in setting priorities and weighing information about treatment options and associated risks or benefits. This power to mediate and priorize information occurs within a linguistic and cultural "black box" which both the patient and the clinician must access through the interpreter. Within this "black box" the interpreter can actively intervene on the patient's behalf to explain a wider range of treatment priorities or clarify the client's right to refuse. As a hospital employee and paraprofessional health worker, the interpreter-advocate may use his/her cultural knowledge and personal rapport with the patient to reinforce the clinician's definition of appropriate treatment choices and thereby compromise his/her advocacy role.

From medical, legal and sociocultural perspectives, the inclusion of cultural mediators and advocates in consent proceedings raises a series of significant issues. A primary question is what standards should be applied in assessing the impact of language ability and cultural access in determining whether a patient is competent to make decisions. When intermediaries are introduced to facilitate communication, their impact upon the general clinician-patient relationship must also be evaluated. Where members of the family or other non-professionals are used as interpreters, the potential impact of their values and personal agendas should be considered. The level of control exerted by both professional and informal interpreters through selective translation, exclusion and "weighing" of information requires evaluation. It may become necessary to determine whether intermediaries assume some legal or ethical responsibility for decisions.

In summary, sociocultural analysis of real interaction sequences in the negotiation of consent between clinician and client differs fundamentally from legal or ethical analysis of the marker decisions. For Native clients, agreements may reflect the emergence of trust relationships achieved through an extended, incremental process of exchange rather than a formal, final contract. Interactionist and ethnomedical approaches more clearly reveal the communication processes and power relations which are part of the process of translating and priorizing information. Our analysis of the role of Native medical interpreters in both case studies clearly indicates that dyadic clinician-client interaction is strongly influenced by intermediaries. Both confirm that translators, cultural brokers and personal advocates negotiate shared meanings and influence the balance of power in cross-cultural, clinical communication. As well as demarcating formal legal and ethical decision points, cross-cultural consent agreements also function as integrative rituals through which participants reconcile power imbalance and negotiate clinical trust.

ACKNOWLEDGEMENTS

The authors would like to acknowledge the special contribution of our colleagues and research associates including William Koolage, Margaret Smith, Ellen Haroun, Andrew Koster, and Charlene Ball. We also wish to thank Drs. Pat Kaufert, Barney Sneiderman, John Walker, Gareth Williams, and Brian Postl for their editorial assistance. Our initial study of hospital-based interpreters was financed by a grant from the Manitoba Health Research Foundation. Subsequent research on interpretation and health communication in Inuit communities was financed by a grant from National Health Research Programs Directorate of Health and Welfare Canada (Project No. 6607-1305-49).

REFERENCES

Barber, Bernard
    1980   Informed Consent in Medical Therapy and Research. New Brunswick, New Jersey: Rutgers University Press.
Baskett, T.
    1978   Obstetric Care in the central Canadian Arctic. British Medical Journal 2: 1001-1004.
Culhane Speck, D.
    1987   An Error in Judgement: The Politics of Medical Care in An Indian/White Community. Vancouver: Talonbooks.
Faulder, C.
    1985   Whose Body Is It? The Troubling Case of Informed Consent. London: Virago Press.
Fletcher, J.
    1983   The Evolution of Ethics in Informed Consent. In K. Beg and K.E. Tranoy (eds.) Research Ethics. New York: Alan R. Liss. Pp. 187-228.
Freidson, E.
    1970   Professional Dominance: The Social Structure of Medical Care. New York: Atherton.
Fox, R.
    1980   The Evolution of Medical Uncertainty. Milbank Memorial Fund Quarterly, Health and Society 58: 1-49.
Goffman, E.
    1967   Interaction Ritual: Essays in Face-to-Face Behaviour. Chicago: Aldine.
Good, B. and Good, M.J.
    1981   The Meaning of Symptoms. In L. Eisenberg and A. Kleinman (eds.) The Relevance of Social Science for Medicine. Dordrecht: D. Reidel. Pp. 1-26.
Gray, B.H.
    1978   Complexities of Informed Consent. Annals of the American Academy of Political and Social Science 437: 37-48.

Hahn, R. and Gaines, A. (eds.)
    1985   Physicians of Western Medicine: Anthropological Approaches to Theory and Practice.
           Dordrecht: D. Reidel.
Harwood, A. (ed.)
    1981   Ethnicity and Medical Care. Cambridge: Harvard University Press.
Helman, C.G.
    1984   Culture, Health and Illness: An Introduction for Health Professionals. London: Wright.
    1985   Communication in Primary Care: The Role of Patient and Practitioner Explanatory Models.
           Social Science and Medicine 20: 923-931.
Hopp V. Lepp
    1980   112 D.L.R. (3d) 67, 13 C.C.L.T. 66, (1980) 2 S.C.R. 192.
Jakobovits, I.
    1983   The Doctor's Duty to Heal and the Patient's Consent in Jewish Tradition. In G. Danston and
           M. Seller (eds.) Consent in Medicine. London: King Edward's Hospital Fund. Pp. 32-37.
Katon, W. and Kleinman, A.
    1981   Doctor-Patient Negotiation and other Social Science Strategies in Patient Care. In L.
           Eisenberg and A. Kleinman (eds.) The Relevance of Social Science for Medicine.
           Dordrecht: D. Reidel. Pp. 253-282.
Kaufert, J. and O'Neil, J.
    1988   Cultural Mediation of Dying and Grieving Among Native Patients in Urban Hospitals.
           Paper Presented at SSHRC Conference on Death and Grieving. McMaster University, May
           6-7.
Kaufert, J. and Koolage, W.
    1984   Role Conflict Among Culture Brokers: The Experience of Native Canadian Medical
           Interpreters. Social Science and Medicine 18: 283-286.
Kaufert, J., O'Neil, J. and Koolage, W.
    1986   Culture Brokerage and Advocacy in Urban Hospitals: The Impact of Native Language
           Interpreters. Santé, Culture, Health 3, 2: 2-9.
Kaufert, J. et al.
    1984   The Use of "Trouble Case" Examples in Teaching the Impact of Sociocultural and Political
           Factors in Clinical Communication. Medical Anthropology 8: 36-45.
Kleinman, A.
    1980   Patients and Healers in the Context of Culture. Berkeley: University of California Press.
    1975   Explanatory Models in Health Care Relationships. In National Council for International
           Health: Health of the Family. Washington, D.C.: Council for International Health. Pp.
           159-172.
Lautt, M.
    1987   Problems of Applying the Laws on Informed Consent: The Case of The Native Patient.
           Unpublished manuscript: Issues of Law and Bioethics, Faculty of Law, University of
           Manitoba.
Lazarus, E.
    1988   Theoretical Considerations for the Study of the Doctor-Patient Relationship. Medical
           Anthropology Quarterly 2: 34-59.
Lechat, R.
    1976/77 Intensive Sterilization for the Inuit. Eskimo 12: 5-7.
Matthews, J.
    1983   The Communication Process in Clinical Settings. Social Science and Medicine 17:
           1371-1378.
Meisel, A., Roth, L. and Lidz, C.
    1977   Toward a Model of the Legal Doctrine of Informed Consent. American Journal of
           Psychiatry 134: 285-289.

Meisel, A. and Roth, L.H.
    1980   What We Do and Do Not Know About Informed Consent: An Overview of Empirical
            Studies Paper presented at Annual Meetings of The American Psychiatric Association, San
            Francisco, May 8.
Natanson, V. Kline
    1960   354 P. 2d 670.
O'Neil, J.
    1988   Self-determination, Medical Ideology and Health Services in Inuit Communities. *In* G.
            Dacks and K. Coates (eds.) Northern Communities: The Prospects for Empowerment.
            Edmonton: Boreal Institute. Pp. 32-50.
Reibl V. Hughes
    1980   114 D.L.R. (3d) 1, 14 C.C.L.T. 1, (1980) 2S.C.R. 880.
Scheper-Hughes, N. and Lock, M.
    1987   The Mindful Body: A Prolegomenon to Future Work in Medical Anthropology. Medical
            Anthropology Quarterly 1: 6-41.
Sharpe, G.
    1979   Options on Medical Consent. A Discussion Paper Prepared by Ontario International
            Committee on Medical Consent. Ontario Ministry of Health.
Somerville, M.A.
    1981   Structuring the Issues in Informed Consent. McGill Law Journal 26: 740-754.
Stimson, G. and Webb, B.
    1975   Going to See the Doctor. London: Routledge and Kegan Paul.
Taylor, K.M. and Kelner, M.
    1987   Informed Consent: The Physican's Perspective. Social Science and Medicine 24: 135-143.
Taussig, M.
    1980   Reification and Consciousness of the Patient. Social Science and Medicine 14B: 3-13.
Tuckett, D. and Williams, A.
    1984   Approaches to the Measurement of Explanation and Information-Giving in Medical
            Consultation: A Review of Empirical Studies. Social Science and Medicine 18: 571-580.
Turner, V.W.
    1968   The Drums of Affliction: A Study of Religious Processes among the Ndembu of Zambia.
            Oxford: Clarendon Press.
Waitzkin, H. and Stoeckle, J.
    1976   Information Control and the Micropolitics of Health Care: Summary of an Ongoing
            Research Project. Social Science and Medicine 10: 263-76.
White, W.D.
    1983   Informed Consent: Ambiguity in Theory and Practice. Journal of Health Politics, Policy
            and Law 8: 99-119.
Young, A.
    1982   The Anthropology of Illness and Sickness. Annual Review of Anthropology 11: 257-285.
Zola, I.K.
    1981   Structural Constraints in the Doctor-Patient Relationship: The Case of Non-Compliance. *In*
            L. Eisenberg and A. Kleinman (eds.) The Relevance of Social Science for Medicine.
            Dordrecht: D. Reidel. Pp. 241-253.

ALLAN YOUNG

MORAL CONFLICTS IN A PSYCHIATRIC HOSPITAL TREATING COMBAT-RELATED POSTTRAUMATIC STRESS DISORDER (PTSD)

This paper is divided into two parts: a description of the topography of ethical judgement in psychiatric settings is followed by an ethnographic account of some ethical dilemmas and disputes associated with the treatment of combat-related posttraumatic stress disorder (PTSD).

ETHICS

As I use the term here, "ethics" refers to a hierarchy of rules: at the base, concrete rules describing the behaviour that a person or category of people owes to other people (or to himself) under certain circumstances (e.g. medical ethics, professional ethics, ethics of self-perfection); at the top, principles (e.g. utilitarian ethics) that are used to evaluate all forms of social behaviour, including behaviour guided by concrete ethical rules.

Ethics, both concrete rules and principles, consist in part of normative codes: social obligations that are backed by formal and informal sanctions. Ethics also include rules and principles which, once breached, are followed by feelings (particularly guilt, shame, dysphoria, and anger) which, like the sanctions associated with norm-ethics, give people reasons for conforming to ethical standards. In practice, norm-ethics and feeling-ethics often coincide, so that people are doubly motivated to behave in prescribed ways, i.e. they want to avoid punishment or opprobrium and, at the same time, want to avoid feeling guilty or ashamed. But there are also situations where sets of norm-ethics have no counterpart feeling-ethics – in which case, a person has no reason to behave ethically if he thinks he can evade detection or punishment. Similarly, there are situations where feeling-ethics have no matching norm-ethics, e.g. because a person's rules are idiosyncratic.

By this definition, ethics are a subset of moral beliefs. "Morality," as it is conventionally understood, also includes ritual obligations and behaviour owed to spiritual beings. In the minds of some people, these obligations should be distinguished from "ethics," which are limited to relations between people. But this is not a subjectively salient distinction in the psychiatric institution described here, and therefore I feel free to use the terms "ethical" and "moral" interchangeably.

Ethics do a variety of jobs. They are guides to action, telling people what to do next and what they can expect other people to do in certain situations. People also use ethics for building narrative accounts of social encounters and events, for shaping arguments, and for persuading other people to make certain choices or to form certain impressions. At the same time that ethics help people to organize

G. Weisz (ed.), Social Science Perspectives on Medical Ethics, 65–82.
© 1990 *Kluwer Academic Publishers. Printed in the Netherlands.*

action and meanings, there are situations where ethics are obstacles to these same ends – for instance, where people find themselves obligated to follow conflicting norm-ethics, or feeling-ethics conflict with concrete norm-ethics, or ethical choice is mired in conflicting long-term and short-term outcomes. It is conflicts like these that I am focusing on in this paper.

Based on my impressions and observations of how people in Western society talk and write about ethical rules and principles, I believe that everyman's tacit understanding of ethics parallels the way in which ethical ideas are represented in the writings of mainstream moral philosophers and professional ethicists. That is, the distinctions commonly made by ethicists, e.g. between autonomy and beneficence, seem reasonable and salient to most laymen if they are put into everyday language. There is one important difference, though. While ethicists give systematic accounts of ethical principles (see the diagram, below), other people are generally unable to do this. Everyman's ethics are either silently embedded in narratives and behaviour or invoked didactically in bits and pieces, usually during disputes and in homilies for children. In this paper's concluding section, I take a closer look at the differences which separate the ethicist's and everyman's views of ethics.

THE TOPOGRAPHY OF ETHICAL JUDGEMENT

The topography of ethical judgement is represented schematically on the following page. Moral philosophers (and, I am arguing, nearly everyone else) make a basic distinction between decisions based on *deontological* principles, which require a person to do what is right even if this choice leads to an unwanted outcome, and *consequentialist* principles, which require a person to choose the alternative that seems most likely to produce the best effects. Ethical judgments can also be based on a combination of deontological and consequentialist principles, e.g. when an instrumentalist is expected to make his choices consistent with a professional code of behaviour. This kind of mixed principle is called *rule-utility*.

Consequentialists are divided into people whose decisions are based on self-interest, i.e. *agent-centered choices*, and people who are oriented to some *collective interests*. A further distinction is made between consequentialists who expect a person to choose the alternative which brings the greatest *net pleasure* and consequentialists who expect a person to choose the alternative which he thinks is most likely to satisfy a highly *valued desire* which need not be the desire to maximize pleasure.

There are two ways in which desires and ways to satisfy desires are chosen. On the one hand, a person is left free to make his own choices, in which case we say he has *autonomy*. On the other hand, choice is put in the hands of someone who is wise and *beneficent*, i.e. someone who can distinguish between the other person's desires and his best interests. According to some writers, choices based on autonomy and beneficence – what a person wants and what he should want – are likely to converge when the person is fully rational, fully informed, and able to

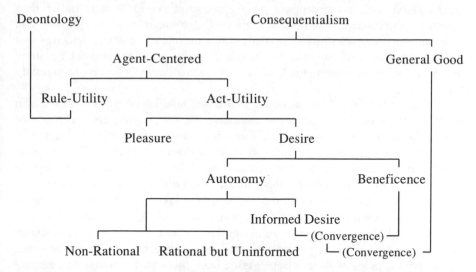

foresee the ultimate effects of his choices. When a person has these attributes, his decisions are said to be based on *informed desire*. Some writers, including J.S. Mill and the young Karl Marx, go a step further and argue that fully informed desire – desire which is no longer alienated or distorted by corrupting influences – will converge not only with individual beneficence, i.e. what is best for this person as an individual, but also with the collective good.

A deontological discourse is equivalent to "ethics" when it concerns a person's . obligations to other people. A consequentialist discourse becomes "ethics" when it says things about justifying autonomy, identifying beneficence, and demonstrating convergence between autonomy and beneficence and between individual and collective interests. The distinction between norm-ethics and feeling-ethics cross cuts the distinction between deontological and consequentialist principles, i.e. a person can conform to either principle for either kind of motive.

RATIONALITY AND ETHICAL JUDGEMENT

Everyday ethics in the West are dominated by agent-centered consequentialist principles, rooted in an historically determined notion of "self." It is this particular notion, this commonsense self, and not simply the priority given to consequentialism, that makes Western ethics distinctive.

The Western commonsense self is distinguished by the following combination of features. Self is identified with the seat of consciousness (mind) which is embedded in a physical substratum (brain) and organized around a rational center. While outward and subjective expressions of self/mind (e.g. personality) distinguish one person from another, mind's rational center is timeless and universal, i.e. untouched by history or culture. The center develops and decays, it is weakened by disease and distorted by emotion, but, unlike other parts of the self, it never metamorphizes into something different. The rational center is also clear-sighted, i.e. its perceptions of the material world reflect, rather than interpret, objects and events. The center is also capable of recognizing and reflecting on elements in its mental environment, including desires/motives and feelings such as guilt, shame, and anger. It is because the rational center is universal, timeless, clear-sighted, and reflexive that people's minds are mutually intelligible.

It is belief in this self and its rational core that justifies people's confidence in an ethic based on informed desire. Because the self is inherently capable of making rational assessments and decisions, it is reasonable to believe that people can make ethical judgments in the absence of an absolutist morality. The key term here is "rationality"; this is what makes it possible to leap from deontology to consequentialism. Despite its importance, however, the precise meaning of rationality remains unclear in the discourse of both laymen and experts. It is most commonly represented as having something to do with the coherence of beliefs (e.g. rational discourse is not self-contradictory) and with the consistency between people's beliefs, perceptions, reasons, and actions, and the real and perceived outcomes of their actions. Aside from these abstractions, the meaning of rationality is mainly tacit and ostensive, i.e. embedded in examples of verbal and non-verbal behaviour rather than explicit standards. To an important extent, we know what rationality *is* because we can point to what it *is not*. Non-rationality marks the edges of rationality, and an agent-centered consequentialist ethic which lacked recognizable examples of non-rationality would be an ethic based on an unbounded (edgeless) and therefore useless conception of rationality.

To summarize, rationality is connected to ethics along the following route: without examples of non-rationality, there is no rationality; without rationality, there is no informed desire; if there is no informed desire, there can be no hope for convergence between autonomy and beneficence (between what we *do* want and what we *should* want); and without convergence, there is no justification for an individualistic social ethic.

PSYCHIATRY, RATIONALITY, AND ETHICS

Psychiatry and ethics intersect at (at least) three points. *First*, psychiatric medicine provides institutional surfaces on which conventional ideas about rationality, non-rationality, and the nature of the self are objectified, made tangible and real in clinical experience and given a presence in popular and scholarly discourse. In this way, an historically determined and revisable conception of self is re-presented as an irresistible fact of nature. *Second*, psychiatric medicine establishes an autonomous system of meanings organized around the idea that what is normal (defined in relation to antinomies of health and illness) is coequal or superior to what is merely normative (defined in relation to antinomies of right and wrong). In this way, psychiatry supplements the ordinary language of ethics with the lexicon of psychodynamics (displacement, projection, acting-out, compulsion, etc.), making it possible for clinicians and writers to retell interpersonal events as intrapsychic relations. *Third*, like other sectors of medicine, psychiatry is associated with ethical problems rooted in conflicting claims to beneficence and autonomy. These problems consist of both ethical *dilemmas* (occasions when a patient or clinician believes he must choose between alternative goals or principles that are equally compelling but lead to contradictory outcomes or courses of action) and ethical *disputes* (encounters between people who are advocating conflicting goals or principles).

The remainder of this section illustrates the intersection of ethics and psychiatry with an example of an ethical dispute which I observed in a psychiatric unit in the U.S. Veterans Administration (VA) Medical System.

A psychiatrist and his patient are locked in a dispute. They share a con-sequentialist ethic, but are unable to agree on what goal (desire) they should be pursuing together. On an earlier occasion, the patient was diagnosed as suffering from a "service-connected" disability, for which he now receives a monthly compensation check. His disorder is a source of some unpleasant symptoms, but it is also his main source of income and an important means of fulfilling his social obligations. He calculates the costs and benefits of having his disorder, and decides that he wants to keep it and his current disability rating (which determines the size of his compensation check). On the other hand, his psychiatrist wants to give top priority to restoring the patient to his pre-morbid state. From the patient's point of view, life without his disorder will be significantly more miserable than life with it. The psychiatrist knows his patient's circumstances and desires, but he is committed to do his duty. He rejects the patient's desires and adopts a principle of rule-utility: he will continue trying to cure the patient and separate him from his source of income. The psychiatrist justifies his choice as follows: (a) because he is a physician and employee of the VA, his first obligation is to reduce his patients' disabilities, and (b) by reducing his patients' disabilities, he makes them more employable and independent, and thus enhances their autonomy.

The dispute boils down to a conflict between the patient's claim of autonomy and the physician's claim of beneficence. As the encounter progresses, the patient becomes angry and abusive. When he is reproached by the psychiatrist, he claims that his behaviour is justified, since it originates in his informed desire to punish people, like the clinician, who are responsible for his current unhappy state. The clinician disagrees, and reasons that the patient's psychiatric diagnosis means he is not fully rational, therefore he is incapable of informed desire.

Given the conception of ethics with which I began this paper, one would have to conclude that this encounter has a distinctively ethical character. On the other hand, most conventional thinkers, including this psychiatrist, would reach a different conclusion. They are likely to see the events in medical terms – a matter of symptoms, psychopathology, diagnosis, etc. – and to think of ethical issues as the by-products of medical problems. From an anthropologist's perspective, their "failure" to recognize the encounter's ethical dimension is an important datum, reflecting three enormous social and cultural advantages which a psychiatrist enjoys over his patient in ethical dispute.

First, the clinician has the institutionally constituted power to frame the dispute in his own terms. By adopting a psychiatric frame of reference, he radically undermines the patient's claim to informed desire. Within this frame, he claims privileged access to the thing on which the patient's autonomy is predicated, his mind, and, so, to the true meaning of the patient's words. Against the clinician's powerful claims, the patient gives only ineffective justifications, based on diffuse notions of justice and natural rights: e.g. "I sacrificed my youth for my country. Now it's my country's time to do something for me."

Second, the psychiatrist uses the institutional division of labour to outflank the patient on a major point. When the patient claims that he will be more miserable without his disorder, the psychiatrist responds first by observing that most of his social and economic problems are caused by his disorder and then by sending him to a "vocational specialist" who is responsible for counselling recovering patients about opportunities for work and retraining.

Third, the patient is powerless to change the venue of the contest. Their dispute is about access to certain desiderata: a diagnosis, disability rating, and pension. Ultimate control over these goods belongs to the institution and its agents, including the clinician.

(In this encounter, the patient is pursuing a non-medical goal, a disability pension. Later in the paper, I want to show that ethical disputes of this sort also occur when both patient and clinician are pursuing medical goals.)

COMBAT-RELATED PTSD

The rest of this paper is about everyday life and ethical judgement in a VA inpatient unit specializing in the treatment of combat-related PTSD. For convenience, I refer to this unit as "the Institute."

Psychiatric nosology used in the United States is encoded in the latest edition of *The Diagnostic and Statistical Manual of Mental Disorders* (DSM-III-R), an official publication of the American Psychiatric Association. According to DSM-III-R, all cases of PTSD originate in a psychologically traumatic event "outside the range of usual human experience." The patient's symptoms are described as consisting of (1) distressful reexperiences of his traumatic event, occurring in dreams, nightmares, intrusive thoughts and images, and (less often) dissociative episodes; (2) behaviour aimed at avoiding these reexperiences, e.g. agoraphobic behaviour; (3) numbing of responsiveness to people and events; and (4) persistent symptoms of arousal, such as outbursts of anger, exaggerated startle response, and difficulty falling asleep and staying asleep.

PTSD shares nearly all of these features with combinations of other disorders: depression, generalized anxiety disorder, adjustment disorder, and phobic disorders. This creates problems in diagnosing the disorder, since most patients with combat-related PTSD also have concurrent psychiatric diagnoses, the most common being depression and generalized anxiety disorder. In other words, it might be difficult to tell a man who has concurrent depression and anxiety disorder from a man who has concurrent depression and anxiety disorder *and* PTSD, if it were not for the fact that PTSD has two features which are distinctive. These are its unusual traumatic event and the patient's symptomatic reexperiences of this event. In practice, only the traumatic event distinguishes PTSD from other disorders. PTSD reexperiences are generally similar to the sorts of ruminations and intrusive thoughts that accompany depression and are disorder-specific to the extent that they incorporate content from traumatic events.

Thus PTSD, unlike most disorders in DSM-III-R, is inseparable from its etiology: some evidence of an unusual traumatic event *must* be adduced before a person can be diagnosed as suffering from PTSD.

Nearly all of the Institute's patients are Vietnam War veterans whose disorders are said to originate in experiences in combat operations. Although there is no diagnosis specific to combat-related PTSD, the Institute's patients are different from most civilian victims of the disorder in the following ways:

One, many of these men experienced a series of what could be unusual traumatic events. The events fall into three categories: unusual violence inflicted on the men, unusual violence inflicted by the men, and scenes of unusual violence and/or grotesqueness witnessed by the men. These experiences took place over a long period, lasting from several months to nearly three years. While the patients' traumatic experiences meet the DSM-III-R criterion of "falling outside the range of usual human experience," they were *not* unusual experiences in the men's military units. Further, when men developed psychiatric symptoms in response to what they saw and did, their reactions (conceivably the onset of PTSD) were treated as routine, and requiring no special medical attention, so long as they did not affect the unit's mission or the men's reliability under fire. In this environment, there was a tendency for men to ignore the moral salience of their war-time behaviour until they returned to civilian life.

Two, unlike many civilian victims of PTSD, these men have a chronic form of the disorder. They have waited an average of 12 years between the onset of PTSD-like symptoms and being diagnosed and treated for PTSD. During the years of waiting, most of the men developed maladaptive ways of coping with their most troubling symptoms: sleep disorders, intrusive images and thoughts, poor impulse control, low self-esteem, suicidal ideas, dysphoria, and anxiety. Many men also have post-war histories of alcohol and drug abuse. Patients and therapists generally account for this as a form of "self-dosing."

This summary of the patients' symptoms diminishes their distress in one important respect. It suggests, quite correctly from the clinician's point of view, that there is a pattern and process to the men's post-war lives. This is not how most patients seem to have experienced those years, however. For them, post-war life was an avalanche of misfortunes – broken marriages, long periods of unemployment, psychiatric hospitalizations, jail time, chronic guilt and depression – without any common cause or over-arching meaning. Only after receiving a PTSD diagnosis did men discover the narrative thread which ran through their lost years.

HOW PATIENTS IMPROVE

The Institute's inpatient ward consists of 13 to 16 patients and a staff of 20 to 22, including a psychiatrist, two clinical psychologists, a physician's assistant, a social worker, nurses, and people known in the VA as "rehabilitation counselors."

Before being admitted to the treatment program, men go through three days of inpatient evaluation and assessment. This is intended to screen out men with severe personality disorders, psychoses, and factitious PTSD.

During the two years that I observed life at the Institute (1986-1987), the treatment program was divided into three phases: one week of orientation; eight weeks of individual and group psychotherapy, and training in "coping skills"; and a two week re-entry phase. Most patients were required to extend the middle phase by a month or more, and it generally took four or more months to complete the treatment program. A majority of men left the program before completing it, however, either for personal reasons or for violating the patient contract. The use of alcohol and non-prescribed drugs was the most common reason for involuntary discharges.

The content of the treatment program has evolved since the Institute opened in 1985. Four assumptions have remained constant, however. (1) During war-time, patients experienced several categories of unusual violence. Only *one* category of violence is responsible for their disorders, however. Each man's PTSD originates in a morally forbidden act, such as torturing and killing prisoners, killing civilians, or killing other Americans. There are patients at the Institute who deny ever committing such acts. Some of these men want to hide their acts because they still fear punishment for war crimes and homicides. Other men are assumed to have repressed their memories of the events and to be incapable of retrieving information without the help of therapy. In either case, a man's silence is regarded as prima facie evidence of his act of commission. (2) PTSD's psychodynamic core is a repetition compulsion. Each patient is psychologically compelled to continually re-enact the behaviour that precipitated his trauma. Compulsive re-enacting continues after he becomes a patient at the Institute. (3) In order to recover, each patient must be made to recall his traumatic event and then disclose it, in detail, during a group psychotherapy session. His narrative is the patient's rosetta stone, since it delineates the order (= repetition compulsion) beneath the disorder of his post-war life. (4) Each patient's pathology erects obstacles to recovery. These obstacles may make it difficult for him to recall and disclose the details of his traumatic event, and to recognize his repetition compulsion and its connection to his symptomatic complaints. (5) Psychoactive drugs play no significant role at the Institute. Patients are allowed to continue to receive anti-depressants, but they are weaned off anti-anxiety drugs, which are said to interfere with the tasks of recalling and "processing" traumatic events.

Therapeutic activity at the Institute consists mainly of talking: talking in individual and group psychotherapy, talking in coping skills groups, talking in treatment assessment sessions, and so on. Talking is simultaneously a therapeutic modality and evidence of a patient's progress.

Significant behavioral changes take place very soon after men enter the Institute: sleep patterns improve, violent impulses become more manageable, thoughts of suicide diminish and disappear, intrusive imagery becomes less troublesome, feelings of depression become less severe, and men feel more in control of their emotions. These improvements fluctuate during the period of hospitalization, but most patients are in much better shape when they leave than when they arrived. It is unclear whether these changes are directly attributable to the treatment program, however. Therapeutic talk may have helped to maintain these rapid changes, but it is clear that the changes stem, in whole or part, from the men's adaptation to life in a highly structured and benign environment. (No outcome studies have yet been undertaken to learn whether patients maintain these changes after leaving the Institute.)

What is gradually transformed over the course of the treatment program is not the patients' non-verbal behaviour, but rather the form and content of their talking, specifically their ability to provide appropriate etiological narratives (= details concerning acts of commission) and accounts of their everyday behaviour (= consistent with the notion of repetition compulsion). The clinical ideology says that these verbal changes are more significant than quick changes in the patients' presenting symptoms. Changes in the content of talk are said to be reliable indicators of important intrapsychic changes, and these are said to presage long term symptomatic changes.

ETHICAL PROBLEMS AT THE INSTITUTE

An ethical *dilemma* is a line of reasoning which leaves the thinker with a set of choices each of which is self-defeating. An ethical *dispute* is an engagement between persons that has the possibility of transforming itself into an interminable zero-sum game. In both events, ethical principles change from being guides to action – enhancing institutional order and interests – into obstacles to action. There are three sources of these ethical problems at the Institute.

One, the clinical ideology requires each patient to admit to himself and to others that his disorder originated in an odious act, and that the act originated in his active intention to do harm. From the patient's point of view, this creates the following problem:

In order to live at peace with myself and to be free from disturbing thoughts and images, I must accept responsibility for my war-time acts. I must no longer block out memories of my role in these acts, nor can I claim to have been duped into them by political and military leaders. But once I accept this terrible responsibility, how can I live with myself? What peace can I know?

Telling a patient that he must disclose his traumatic event engenders two ethical problems for the patient. First, there is an ethical dispute based on the patient's claim to autonomy (= either I have no event, or I have no event I want to think or talk about) and the clinician's conflicting claim to beneficence (= you must disclose your event in order to gain a better life). Second, there is the patient's dilemma over informed consent (= his Catch 22: in order to be fully knowledgeable, I must accept unbearable facts).

Two, clinicians know that it is painful and exhausting work for men to recall their traumatic experiences. Patients often have physical and psychological reactions, including gastro-intestinal distress, headaches, nightmares, intrusive images, anxiety, and depression. Therapists are naturally reluctant to inflict pain on their patients, especially when they have their own doubts about the efficacy of the treatment. A clinician's scruples increase when he is told that his patient was only a boy when he was sent to war, he was unprepared for what he encountered, he was manipulated by cynical men he had been taught to trust, he returned home to an ungrateful and contemptuous public, he continues to be treated as a pariah and is discriminated against by potential employers.

There are two ethical dilemmas here. First, the clinician has doubts concerning his informed desire and capacity for beneficence (= Is he knowledgeable enough to inflict pain?). Second, there is the conflict he experiences between deontological principles on the one hand (= do no harm, do not humiliate, seek justice), and his consequentialist principles on the other (= it is beneficent to do whatever is required to restore the patient to normal social life).

Three, there is the ethical dispute, already described, between a clinician's desire to reduce his patient's disability and the patient's economic interest in keeping his disability.

HOW ETHICAL PROBLEMS ARE MODULATED

Three features of institutional life modulate the impact of ethical problems on the Institute's treatment program and the morale of its clinical staff.

1. The clinician's ethical dilemmas are rooted in his knowledge of (a) the patients' odious acts and (b) the pain he inflicts on patients. Not all therapists are equally disturbed by these dilemmas, however. Some of the Institute's clinicians seem to be relatively indifferent to both the patients' distress and the suffering of their war victims. The therapists' empathy for their patients is further diminished by the men's verbal abuse and their sometimes questionable motives for being hospitalized.

Even therapists who are initially troubled by what they hear and do at the Institute, tend to be less affected by ethical issues as time goes by. There are at least three reasons for this:

One, therapists believe that some patients may be giving spurious accounts of their war-time experiences. While some men have reasons for not disclosing acts which they did commit, other patients have economic and psychological motives for confessing to abominations which they never committed. It is time-consuming and often fruitless to try to obtain documentary evidence that might confirm details in their patients' narratives, and only rarely do therapists attempt to get this information. As a result, therapists are often unsure about the facticity of their patients' disclosures, and therapeutic talk about war-time events floats on a sea of uncertainty. Did the narrated event really occur? Did it occur as the patient described it? If it did occur, did the patient actually participate in it?

Second, there is a tendency for therapists to feel that the dilemma that stems from the patient's narratives is less salient than the dilemma that stems from their own role in inflicting pain. The following line of reasoning, abstracted from therapists' talk during clinical supervisions and meetings, outlines how they say they get themselves off the moral hook:

The quantum of suffering inflicted by therapists on patients in the here-and-now is more real and important than the larger quantum of suffering already endured by the patients' victims. Further, suffering which one might be able to reduce (in the clinic) is more important than the irreparable suffering already endured by the patients' victims. Therefore, the patients' suffering is more significant than their victims'. Concerning suffering inflicted on patients in the clinic, therapists argue (and believe?) that it means that a patient is "getting closer to his issues," i.e. he is on the point of recalling his traumatic events. Put in other words, increased distress is possibly a precursor to recovery, and the loss of a small quantum of happiness in the present is less important than the possibility of obtaining a much greater amount of happiness in the future.

Third, when a therapist has a strong negative response to a patient's narrative, his feelings and behaviour are medicalized, i.e. called "counter-transferences" and "stress-responses," and stigmatized, i.e. called unprofessional, counter-thera-peutic, and inimical to the therapeutic alliance between clinician and patient. In this way, the meaning of the therapist's anger and revulsion is displaced from the patient's physical aggression against his victims (in the past) onto the therapist's psychological aggression against his patient (in the present). The clinician's reluctance to inflict therapeutically necessary pain on his patient is accounted for in the same terms: he is described as having a stress-response, being unprofes-sional, victimizing his patient by withholding therapy, and so on.

2. There are also institutional procedures which limit the importance of ethical problems at the Institute. For patients, the most important procedures are meetings of "multidisciplinary treatment teams," composed of the therapists who are responsible for treating the particular patient. Teams meet regularly to assess each patient's progress through the treatment program. They also meet after a man has violated one of the Institute's rules. Violations include ethical disputes over conflicting claims to autonomy and beneficence. These disputes are re-framed in the psychodynamic parlance of the treatment team as "re-enactment," "acting-out," "resistance" and "pathological blaming."

The treatment team has a schedule of graduated responses for dealing with violations: rule-breakers are first warned, then restricted to the ward, and finally discharged. On each occasion, the patient is told that the team's function is therapeutic, it is not a disciplinary body, it makes no moral judgement on his behaviour now or in the past, and its ultimate purpose is to help him see that this kind of behaviour is rooted in his pathology and that it is self-defeating and counter-therapeutic. The same themes are repeated during group psychotherapy sessions.

3. Ethical problems are also modulated through etiological narratives. These are authorized stories about the origins of PTSD, and they are the basis for much of the clinical talk that takes place between therapists and patients during psycho-

therapy sessions and meetings with the multidisciplinary treatment team, and between the clinical director and the therapists during clinical supervision. The next paragraphs retell the Institute's three most important narratives. Two of them originate in Freud's ideas about instinctive aggression and libidinal attachment and in his distinction between traumatic neuroses and psychoneuroses. The other narrative is based on ideas taken from learning theory. Each narrative is believed by the Institute's staff to retell aspects of the true origins of PTSD's symptoms, and to provide guidelines for making clinical decisions, e.g. helping the therapist choose an appropriate response to a patient's violent acting-out behaviour.

Clinicians tell these narratives about PTSD because they think the stories are therapeutic. But PTSD is simultaneously encoded with symptomatic and moral meanings and, in the case of many diagnostic features (e.g. guilt, anger, painful intrusive thoughts and images), the narratives' medical efficacy is indissociable from their moral efficacy. That is, at the same time that the narratives (ostensibly) explain the origins and appearance of the patients' symptomatic distress, they also explain how to resolve the men's moral distress.

Schematically, each narrative takes patients and therapists through a two-step cure:

First, the narratives say that a patient's sense of opprobrium, his fear of spiritual punishment, and his moral outrage are all elaborations of primal feelings like guilt, shame, and anger. (Thus feeling-ethics are transformed into feelings.) It is only through examining these feelings, and not his ideas about moral behaviour, that the patient (and his therapist) can discover the sources of his distress.

Next, the narratives say that these feelings, as the patient is experiencing them, are actually symptoms – that is, things which originate in intrapsychic (not interpersonal) relations. Thus the proper focus of therapeutic interest is the patient's traumatic *experiences*, located in the timeless space of the primary processes, rather than his traumatic *events*, sited in war-time Indochina. It is because of the patient's symptomatic confusion between the meaning of events (morality) and the meaning of experiences that PTSD deserves to be called "disease of time."

THE NARRATIVE OF SELF AND TIME

The first narrative is about a man (= the patient) who goes on a journey where he loses a valuable part of himself, mind's rational center, and is transformed into something less than fully human. Afterwards, he is forced to live in the past, a prisoner of his experiences. He acquires special skills and knowledge which enable him to transform his experiences into memories. He escapes from the past, regains mind's rational center, and is transformed back into his human form.

According to this narrative, the patient's disorder begins with a war-time event in which his ego (which we can identify with the self's rational center) lost its ability to integrate three unconscious forces: his aggressive drive and its urge to destroy, his libidinal drive and its need for tenderness and longing for attachment, and his super-ego. At this point, he was forced to solve an unsolvable puzzle:

How can I reconcile three conflicting desires? He failed, and in the course of his failure discharged his aggressive impulse in a violent and destructive act of commission.

The patient's traumatic act had two important consequences for him. First, the self split into two part-selves: an aggressor-self organized around destructive impulses and a victim-self organized around pathological feelings of tenderness for victims of aggression, especially for dead comrades. The split was accompanied by the increasing peremptoriness of his aggressive impulses, thus explaining the patient's post-war history of explosive violence. Second, the patient developed a compulsion to repeat his traumatic experience, to re-enact the problem solving behaviour that failed him on the original occasion, in a vain attempt to regain a form of mastery that only whole selves can possess.

In this narrative, splitting is simultaneously a symptom of PTSD and an obstacle to the self-knowledge and disclosure needed for recovery. It is only as a part-self that the patient can think and talk about his experiences. Accounts given by part-selves are only part-accounts; they tell two different stories about what happened, not complementary parts of the same story. Second, because part-selves, unlike the whole self, are intrinsically bound to emotions, they are incapable of getting outside of the patient's experiences. Anger and pathological tenderness keep him locked inside of these past events. In this way, experiences acquire a quality of timelessness: it is truer to say that the patient's descriptions of his war-time events are his reexperiences of the events rather than his memories of them.

In order to recover, the patient must be made to recall his traumatic event. Knowledge of the event demystifies the sources of symptomatic behaviour, shows him that he is responding to internal, psychic strivings rooted in the past and not, as he incorrectly believes, to provocations in the here-and-now. To recover, he must believe/know and, in an act of will, publicly acknowledge that he is the author of his own traumatic act, that he actively chose to hurt and destroy when he could have chosen to behave otherwise. Only then – having sublimated his aggressive urges and emerged from the timelessness of PTSD – can he reestablish self's authority and its claims to moral sovereignty.

THE NARRATIVE OF SELF AND SURVIVAL

The second narrative provides a solution to the following dilemma: How can I live with myself once I have acknowledged my odious acts? The narrative says that the although patients must accept responsibility for their behaviour in order to recover, there is a way in which they can avoid accepting *moral* responsibility.

Actually, the part-selves give him similar advice. During group psychotherapy, it is common to hear a man's aggressor-self describe his patient as being "no better than an animal" because of what he did in Vietnam. To say that someone is an animal and no longer human, also puts him outside the scope of moral judgement. This is what the part-self implies when he tells his therapist: "You say you understand what I did, but you can't since even I don't understand." It is also common to hear a victim-self describe how he and his comrades were/are exposed and powerless before their enemies: the Vietcong, the North Vietnamese, VA bureaucrats, therapists. Through his powerlessness, the victim-self substitutes his patient for his victims, and gets the patient off the moral hook. "You're looking for victims, are you? Well, you don't have to look far. We're your victims."

The part-selves resolve the patient's ethical dilemma by removing him from the span of moral judgement. But the cost is enormous, since the patient is effectively cut off from the rest of humanity. The narrative of self and survival gives a different solution. Instead of removing the patient from the scope of moral judgement, the narrative transforms the conditions under which moral judgments are made. It says that in the combat zone patients became habituated to putting themselves under the control of a hard-wired survival circuit, located somewhere beyond mind's rational center. Within this circuit, perceptions of danger lead to feelings of danger and frustration, which elicit feelings of anger and aggressiveness, which then explode in acts of life-preserving violence. Each point in the series is fired off automatically. In Vietnam, habituation was not willed, but took place behind the men's backs. Sometimes it resulted in rationally unwanted events, such as killing innocent people, but ultimately it was adaptive. Men whose behaviour did not become habituated in this way perished. Thus, the patients had no real choice, since habituation was not willed and failure to habituate was self-destructive.

After the patient left the combat zone, habituation to violence became maladaptive, in the same way that hypervigilance, fear of crowds, and unusual sleep patterns also became maladaptive once the patient returned home. The patient's problem now is that he behaves as if he were still in the combat zone.

This helps to explain another symptomatic complaint associated with PTSD: persistent and oppressive feelings of guilt. In ordinary circumstances, where the rational center is in control, feelings of guilt are automatically fired off when people behave in violent or abusive ways. Because guilt feelings are distressful, they function to limit the amount of violent or abusive behaviour a normal person is willing to initiate. Patients who are habituated to the survival circuit accumulate excessive amounts of guilt, as a residue of their frequent violent behaviour. These guilt feelings feed the patient's sense of frustration and anger and frustration and anger result in explosive violence. Thus the meaning of guilt is that it keeps patients habituated to the hard-wired circuit. In the narrative of self and survival, guilt loses its signifying power: it no longer points to something bad (acts of commission), it is in itself something bad, i.e. symptomatic and maladaptive.

THE NARRATIVE OF SELF AND BETRAYAL

Unlike the other two narratives, this is one which therapists tell only to one another.

The other narratives tell how PTSD robs every patient of the resources he needs for overcoming obstacles to disclosure and recovery. It is because of this that he must depend on the help of therapists. And there's the rub. When a therapist comes into contact with a patient's anger, threats, verbal abuse, and pathological tenderness, he is drawn into the patient's pathology. Once this happens, the clinician loses his ability to help his patient. A therapist is robbed of healing power through his fear: fear of the patient's anger and potential for violence, fear of his own anger and his urge to hurt the patient, and fear stemming from his projective identification with the patient's aggressor-self.

A therapist can respond to fear by overcoming it, maintaining a strictly professional stance, rigorously analyzing his counter-transferences, and so on. On the other hand, he can respond by adapting to the source of his fear, his patient. In the latter case, he identifies with the patient's victim-self, and becomes the patient's ally and advocate against perceived injustices of the therapeutic regime. Patients often complain that the Institute's rules are arbitrary, rigidly administered, and that they cause the patients avoidable unhappiness. On the other hand, the Institute's position is that its rules do not have to be fair (= honoring the patients' claims to autonomy), only therapeutic (= beneficent). Sometimes, the value of a rule lies in its arbitrariness. Arbitrary rules can provoke. Through provocation, rules can force part-selves to act out under controlled conditions, make themselves visible, and, in this way, nurture the patient's understanding of his disorder.

The narrative of self and betrayal says that this form of adaptation, in which a therapist is seduced by the pathological strivings of his patient's part-selves, is equivalent to "colluding with the patient in his pathology." Collusion leads to a double betrayal. First, the therapist betrays his patient, by consigning him to a psychiatric purgatory where he will forever reexperience his traumatic event. At the same time, by subverting confidence in the need to inflict therapeutically necessary pain, he betrays his own colleagues, by undermining their faith in the rules, alliances, and narratives on which the therapeutic regime is based.

At the end of the story, the therapist himself re-enacts the morally forbidden act. The flow of time is reversed, and the present floods back into the past.

CONCLUSION: THE PERSON AND ITS SELVES

I return to two points made earlier in this paper. One, because of the way in which combat-related PTSD is clinically constructed at the Institute, the *therapeutic* efficacy of its treatment program (the ability to control or eliminate symptomatic distress) is indissociable from its *moral* efficacy (the ability to undo ethical disputes and dilemmas). Two, the actual efficacy of the treatment program is problematic. Because most improvements occur soon after patients enter the Institute, one can reasonably argue that these changes reflect the patients' strategic adaptation to the Institute's system of routines, sanctions, and rewards, rather than mental transformations attributable to psychotherapy. There is an important exception to this generalization, however. The form and substance of patients' talk during clinical sessions does tend to change gradually and in clinically appropriate ways, consistent with the idea that the men are responding to the treatment program. For example, although nearly all patients are initially reluctant to talk about painful war-time experiences or to psychologize their conflicts with the Institute's staff, most men eventually disclose their traumatic events during group psychotherapy.

But even here, in the case of verbal behaviour, the meaning of the changes is equivocal. We need to know whether changes in what a man says also signify changes in the mental apparatus which, according to the staff, is the locus of his disorder. Regrettably, there is no way to read mental structures directly from verbal behaviour, and no way to see whether a pathologically split self is reintegrating itself. If we judge a patient only on his words, the most we can say is that he sounds *as if* he is improving. But even if we could go from a man's words into his mind, things would remain unclear, since the Institute's patients often give contradictory narratives. It is not unusual for a patient, during group psychotherapy, to describe his present behaviour in terms of the repetition compulsion thesis; then, a couple of hours later in the day-room, away from the staff, he tells a group of patients that he thinks there is no psychological connection between his behaviour and his traumatic event.

The Institute's therapists, who realize that patients sometimes give conflicting accounts, explain their contradictions in terms of the following possibilities. (1) The patient is rational, believes either A (e.g. the repetition compulsion thesis) or not-A but not both. Therefore he is lying to someone. Either he is making up stories for the staff or he is embarrassed by his conversion to the staff's psychodynamic views and is lying to the other patients. (2) The patient is rational, but he also believes both A and not-A. He recognizes the contradiction however, and is bothered by "cognitive dissonance," originating in the fact that self/mind does not normally tolerate contradictions. (3) The patient is not fully rational, and a split self continues to influence his perceptions and behaviour. He simultaneously accepts and rejects the staff's explanations, without being troubled by the contradiction.

It is because the therapists tacitly identify a healthy mind with the common-sense self that they think a patient's conflicting accounts of his behaviour are necessarily a "problem" needing a solution, and that these three possibilities are the best (most logical) way to frame the problem and its solution. This line of reasoning conveniently ignores the fact that the third possibility (i.e. self-contradiction without distress) is not particular to people in mental hospitals. Mentally competent people regularly produce untroubled contradictions.

In this connection, a number of philosophers have commented on the failure of conventional ideas about the self to explain the pervasiveness of certain mental habits, notably self-deception (How can the unitary self hide things from itself or fool itself?), akrasia (Why would a rational and unitary self fail to choose what it believes to be its best course action?), and other forms of irrationality and lapses from consistency (Davidson 1982; A. Rorty 1980 and 1985; R. Rorty 1987). To see how the philosophers' arguments throw light on life at the Institute, we need to go back to the page where I said that it is important to make a clear distinction between how ethicists think about ethics (i.e. what a person can/should know, want, and do in his relations with other people) and how ethics are represented in the rhetoric of everyday life. My point was that ethicists think of ethics as systems of ideas, analogous to grammars. On the other hand, when everyman talks about ethics and moral obligations, his ideas are generally unsystematized and embedded in narratives, images, and emotions. In lieu of a grammar of ethics, everyman works out of a library of narratives and a gallery of images.

Western ethical discourse is dominated by an agent-centered consequentialism, in which the burden of moral responsibility is conceived in terms of informed desire, beneficence, etc. This burden is justified (i.e. it seems reasonable) because the moral agent is identified with his mind, this mind is identified with the commonsense self, and the commonsense self is believed to possess the apposite qualities and capacities: autonomy, clear-sightedness, etc. But why assume that everyman's ethics also require this self? Why suppose that his ethics, which are rooted in narratives, images, and emotions, require a fixed and unitary subject? To the contrary, everyday talk gives the impression that consciousness apprehends itself through a number of selves (or quasi-selves, if you like), each configured in a set of narratives and images, and each constituting the subject of the person's experiences at some point in time. Further, everyman seems generally untroubled by the inconsistencies among his narrative selves, at least until they are called to his attention and he is made (implicitly) responsible for explaining them.

The idea which I am advancing is that every person *ordinarily* encompasses several selves, each of which is equivalent to a more or less coherent ensemble of beliefs, perceptions, and feelings. Thus everyman's multiple selves are unlike the pared-down selves that psychiatry and theology create from split-off emotions and dichotomous dispositions (e.g. Eros versus Thanatos at the Institute).

The idea that each person encompasses several selves, spanning past and present, is not an argument against the existence of the commonsense self. Rather, the idea is to reposition this self. Clearly the commonsense self does exist, since we hear it every day, talking through the lips of patients, therapists, and ourselves. Indeed it is partly because of the influence of this self over daily life that people feel the obligation to account for inconsistencies among their narratives. My point is that the existence of the commonsense self needs to be accepted, but not on its own terms. It needs to be listened to, but as one voice in a crowd of voices – albeit the voice which, in many people, is permitted to shout longer and louder than the rest.

This, then, is the way that philosophers solve the "problem" of everyday self-deception, akrasia, and irrationality: by putting the commonsense self in its place, alongside the person's other selves. At first sight, the solution may seem to rob ethics of moral agency and open the door to unprincipled moral relativism, making selves rather than people accountable for behaviour. The idea of multiple selves does not necessarily point in this direction however, since a person's selves are not autonomous. They are embedded in something bigger: a wider, looser, and centerless configuration of habits of thinking (including the habit of rational calculation), perception, motivation, and action. The selves are morally important because they give reasons for and meaning to a person's purposeful acts. But it is the embodied person and not any particular self that is the locus of moral responsibility.

Thus the idea of multiple selves makes the moral life of the Institute something different from and more complicated than what commonsense suggests. It makes the meaning of a person's life events an inalienable part of his or her selves, not some kind of mental trace that can be siphoned off in the form of symptoms. It socializes mind's rational core, pointing us in the direction of the social institutions and cultural practices which encourage, and often exploit, everyman's mistaken faith in the unity of mind and the autonomy and clear-sightedness of the commonsense self. Finally, it provides a reasoned argument for rejecting the premises and claims of reductionist psychologies and totalizing psychotherapies. This is not the least of its attractions.

ACKNOWLEDGEMENTS

I want to thank Paul Emery, Jack Smith, and Glenn Davis, without whose help I would have been unable to undertake the field research on which this paper is based. I am also grateful for the material support provided to me by the U.S. Veterans Administration.

REFERENCES

Davidson, Donald
    1982   Paradoxes of Rationality. *In* R. Wollheim and J. Hopkins (eds.) Philosophical Essays on Freud. Cambridge: Cambridge University Press. Pp. 289-305.
Rorty, Amelie O.
    1980   Akrasia and Conflict. Inquiry 23:193-212.
    1985   Self-deception, *Akrasia*, and Irrationality. *In* J. Elster (ed.) The Multiple Self. Cambridge: Cambridge University Press. Pp. 115-131.
Rorty, Richard
    1987   Freud and Moral Reflection. *In* J. Smith and W. Kerrigan (eds.) Pragmatism's Freud: The Moral Disposition of Psychoanalysis. Baltimore: Johns Hopkins Press. Pp. 1-27.

CHARLES W. LIDZ AND EDWARD P. MULVEY

INSTITUTIONAL FACTORS AFFECTING PSYCHIATRIC ADMISSION AND COMMITMENT DECISIONS

Discussions about the ethical and policy implications of health care financing are increasingly common in the medical and policy literature. Recent articles in major journals have reported both that current efforts at cost control are having no effect (Callahan 1988) and that the efforts at cost control are seriously undermining the quality of health care (Shortell and Hughes 1988). Whether both of these, one of these, or neither of these findings are correct, we have all become acutely aware of the fact that health care of one sort or another takes up an increasing proportion of the Gross National Product and will probably continue to do so in the near future. This is clearly an extremely complex policy issue that will not soon leave us.

The debate over the effectiveness of various means of cost control has been accompanied by a smaller but growing discussion in the ethical literature about how the individual health care practitioner ought to take such issues into account when making clinical decisions. The core of the debate concerns how the individual clinician should deal with the financial pressures to make decisions that run counter to the individual patient's interest. This literature reflects an agreement that the clinician has the obligation to provide the best care possible within the limits imposed by the need to make prudent use of financial resources. This general ethic, however, rarely helps in practice. Beauchamp and McCullough state the dilemma nicely:

> Under the beneficence model, the physician has an obligation to ... [continue treating the patient], at least until it becomes clear that it is not in [the patient's] best interests to do so. [Yet] the physician also has obligations to the hospital, which in turn has obligations to others to use its resources efficiently.... There is no tidy resolution ... (Beauchamp and McCullough 1984: 147).

Indeed, clinicians increasingly find themselves in situations in which their commitment to providing the best care for their patients conflicts with the hospital administration's desires to control costs or improve revenue. While there is some question in the literature about the appropriate response, most ethicists seem to expect that clinicians will favor treating the patient as well as possible. For example, the President's Commission report on Securing Access to Health Care states:

> sole reliance on professional judgment in setting limits [in the use of resources] is not appropriate because of professionals' tendency to provide all possible medically beneficial care (President's Commission for the Study of Ethical Problems in Medicine and Biomedical and Behavioral Research 1983: 39).

G. Weisz (ed.), Social Science Perspectives on Medical Ethics, 83–96.
© 1990 Kluwer Academic Publishers. Printed in the Netherlands.

Most ethicists seem to see this as the proper response as well. Beauchamp and Childress insist that physicians ought to put the patient first.

> Physicians are to do all they can for their patients without taking into account the kinds of factors ... that policy makers rightly should consider (Beauchamp and Childress 1983: 213).

Even many supporters of cost control have insisted that such decisions should not be made by the physician but should be left to others (Veatch 1985: 14, 1986: 32; Levinsky 1984).

Yet there is some opposition to this position (Thurow 1984). Morriem has argued that the conflict is ultimately not between patient and hospital. Instead the issue:

> pits the interests of the current patient against those of other patients, especially future patients whose health care will one day suffer if hospitals are fewer in number and less well maintained.... Physicians cannot ignore the welfare of their own hospital if they wish to ensure that high-quality facilities are available for treating their patients (Morriem 1985: 32).

In this paper we want to take a somewhat different approach to these issues. We want to ask how clinical norms impinge on the clinician's behavior when faced with issues relevant to the financing of care and vice versa. Given that the publicly acknowledged norms concerning physician decision-making require the physician to put the interests of the patient first, how does the physician deal with conflicting pressures? Does the physician's perceived "best treatment" get sacrificed to cost containment or other financial management concerns?

METHODS AND SETTING

The data for this paper are drawn from a larger study concerning the clinical assessment and management of psychiatric patients. The study was done at a 200 bed urban non-profit psychiatric hospital which primarily provided services to a geographically defined catchment area. Eighteen observers were trained in observational methods, speedwriting, the elements of psychiatry, and the basic issues in the legal regulation of psychiatry. They were then assigned to cover all three shifts of the hospital's emergency room and the day shifts of all of the adult non-geriatric units of the hospital. This allowed them to observe interactions in these settings and to produce a nearly verbatim text of almost all of the conversations between patients, patients' families and the staff in the emergency room as well as the morning rounds, post-rounds reviews, team meetings and morning nursing reports on the units. In all, the observers recorded 417 admission or referral decisions concerning 390 unique patients and 222 episodes of hospitalization.

The observers' notes were dictated and transcribed into a computer along with abstracts of the clinical records and copies of the case summaries written by the clinicians who saw the patients. The notes were then closely coded for a wide variety of issues relating to treatment decision-making. This sort of close coding of the text, in combination with some creative use of a relational data base manager permits the combining of qualitative and quantitative approaches in examining the databases. This was facilitated by the development of a computer program for rapid retrieval of any specifiable sections of the text and the related ability of this program to review those portions of the text in the context of the rest of the case material. The database can be examined both by looking at specific subgroups of interest and, more generally, by looking at various combinations of quantitative and textual data.

The institutional setting chosen raises special financial issues. The financial pressures in a psychiatric emergency room are quite different from those found in a general medical emergency room since psychiatry is not currently reimbursed through the DRG system.[1] However in psychiatry as elsewhere, the clinicians find themselves directly faced with conflicting obligations to what they perceive as the patient's best interests and to the financial interests of the hospital which pays their salaries. The basic concern of the hospital administration to maximize the financial return on the care that the staff provides was consistently and clearly communicated to the staff. In this paper we will deal with two specific financial issues, the administration's concern to keep the hospital as full as possible, (the census issue) and its desire to admit only paying patients (the insurance issue).

Several things must be noted about the way in which care was financed in the hospital that we studied. First, the policy of the hospital was to keep itself at or close to capacity census at all times. While the formal capacity of the hospital was set by a regulatory body, the same body allowed the hospital to maintain a limited number of emergency over capacity beds. This allowed the hospital, during the period of our observation, to function with a mean census of 130 for adult beds when the formal capacity was 133 for adult beds. The problem of coping with a rash of admissions was dealt with by simply admitting the patients to over census beds. Unfortunately on some occasions patients showed up in the ER when the adult beds of the hospital were as much as 8 patients over census.

The management of the hospital census was a continual concern for the hospital administration. At one point the problem of not filling bed space was serious enough that clinical staff were called to a meeting with several high-level administrators who explained in some detail how the ability of the hospital to support various extra benefits for the staff depended on keeping the hospital as close to full as possible. Likewise, during the period of our observations, when the census got too low one of the medical administrators wrote on a chalk board in the ER staff room, in large capital letters, THINK ADMISSION.

The second feature of the setting that needs discussion concerns the reimbursement structure. While there are a variety of different ways in which reimbursement is managed, from the point of view of the hospital only three special cases are important. First, the Commercial Insurance patients generally have all of the hospital charges paid. This means that the hospital was reimbursed about 80% more for these patients than for most other patients, such as those on Medicare or Medicaid who were paid for at a lower fixed rate of reimbursement. Second, the small numbers of HMO[2] patients were usually (but not always) transferred to another hospital with which the HMO had a contract for inpatient care or to the HMO itself for outpatient care. In emergency situations the HMO sometimes agreed to pay the hospital for short term care. Finally, the county agreed to pay for those patients who were from the hospital's catchment area and who had no means of paying for themselves. However this payment came from a fixed pot from which the county paid all hospitals. Each hospital was assigned its own allocation of money at the beginning of the year from which that hospital was to be paid. If the pot was overdrawn during the year, the particular overdrawing hospital was to be compensated for its expenses from other hospitals' allocations. Unfortunately all hospitals typically used their allocations by the middle of the year and all such patients became economically losing propositions for all of the hospitals. It is also worth noting that, since charges at this hospital were over $1000 per day exclusive of any sort of therapy, few people could pay for their own treatment. Table 1 shows the percentages of patients seen in the emergency room with different forms of insurance.

Table 1: Distributions of Insurance Types in Sample

	%	N
Blue Cross/Blue Shield	23.5	99
Commercial Insurance*	3.6	15
Medicare**	5.9	25
Medicaid	50.5	213
HMO	1.2	5
County Insurance	10.2	43
None	5.0	22
	99.9%	422

*   Multiple insurances were counted only once by including them only if not included in a previous category in this list.
**   No patients under 14 or over 59 were included in this sample accounting for the small group of medicare patients.

THE PROBLEM

The empirical question at issue is that of the effect of these two factors on decisions in the Emergency Room. The purpose of this paper is to examine the impact of the administration's concern to keep the hospital full (the census issue) and to admit paying patients (the insurance issue) on dispositional outcomes.

There are two obvious possibilities. First clinicians may make clinical decisions about admission and involuntary commitment so as to maintain a maximum census of fully reimbursed patients. They might do this because they are interested primarily in the welfare of the institution and thus their own self interest and/or the welfare of their future patients or because they are intimidated by the administrators. Alternatively, because they function under a traditional model of beneficence, they might simply ignore these issues.

On the basis of the sociological literature, we were uncertain what we would find. One of the classical debates in medical sociology concerns Talcott Parsons' (1951) contention that the medical profession is committed to a normative system with large components of "collectivity orientation" in which physicians put the interests of their patients before their own. In contrast Freidson (1970) and other critics have contended that such altruism is primarily rhetorical. Both sides have been able to marshall evidence to support their positions. In many ways this paper addresses this issue again: to what degree do health care professionals put what they perceive as in the patient's best interest before other concerns?

From the discussions in the emergency room we felt certain that the census and insurance factors played significant but not overwhelming roles in the decisions. Certainly the clinicians with whom we talked repeatedly express concern about the pressures that were brought to bear on them to consider economics in the care of patients. Indeed, although morale in the hospital was fairly good, we several times heard the hospital analogized to a house of prostitution with the associated metaphors about what was being done to patients and staff. The focus of these expressed concerns was invariably financial pressures.

FINDINGS

In order to assess the impact of these external variables on clinical decisions in the emergency room, we decided to look at the relationships between two independent and two dependent variables connected with each of the 417 patient appearances in the emergency room. The independent variables were census at the time of each patient's presentation in the emergency room (divided into those one standard deviation above the mean, those one standard deviation below the mean and the two-thirds in the middle) and the patient's insurance status. The dependent variables were the patient's disposition from the emergency room (admitted, transferred or referred to another facility, and referred for outpatient treatment) and whether or not the patient was committed on an emergency basis. For this purpose we removed all patients who were brought to the emergency room on a long term commitment since the emergency room staff perceived it as beyond their authority to decide about whether or not to honor these commitments.

Table 2 shows the relationship between census and legal status. Since emergency commitments are one of the easiest ways in which a hospital can increase its admissions, we thought it possible that we would see some inclination to commit patients when the census was low. There is no evidence of that. There

Table 2: Legal Status by Census

| Census Groupings | Legal Status of Commitment | | Row Total |
	Voluntary	Involuntary	
Low census	54	12	66
			15.9
Medium census	218	64	282
			68.1
High census	58	8	66
			15.9
Column Total	330	84	414
	79.7	20.3	100.0

Chi-Square	D.F.	Significance
3.91304	2	0.1413

Lambda: (Asymmetric with Legal Status Dependent) = 0.0000

is a very slight tendency to not commit when the census is high but the relationship is not statistically significant and may well be random fluctuation. In any case, this shows only a possible tendency not to admit when the hospital is over full, not a tendency to use the legal power of commitment to hospitalize patients in order to fill up the hospital.

Table 3 looks at the effect of insurance status on the commitment. Once again there is no tendency to commit the patient based on whether or not, or how, the patient is insured.

Table 3: Legal Status by Insurance Status

| Insured | Legal Status of Commitment | | Row Total |
	Voluntary	Involuntary	
No	19	5	24
			5.8
Yes	311	79	390
			94.2
Column Total	330	84	414
	79.7	20.3	100.0

Chi-Square	D.F.	Significance
0.00465	1	0.9456

Lambda: (Asymmetric with Legal Status Dependent) = 0.0000

Table 4: Disposition by Census

Census Groupings	Inpatient	Transferred or Referred	Outpatient	Row Total
Low census	32	10	22	64
				15.9
Medium census	157	67	52	276
				68.5
High census	27	22	14	63
				15.6
Column Total	216	99	88	403
	53.6	24.6	21.8	100.0

Chi-Square	D.F.	Significance
12.65505	4	0.0131

Lambda: (Asymmetric with Disposition Dependent) - 0.0000

Table 4 shows the test of what seemed a more likely hypothesis, that the frequency of inpatient admission versus transfers or outpatient treatment increases when the hospital is relatively empty. The chi-square statistic for this table is significant showing that these two categorical variables are associated beyond what would be expected by chance. Further examination, however, makes this result appear unimpressive. First, the large sample size is probably largely responsible for the statistical significance of the chi-square, especially when one considers that the lambda statistic (a measure of predictive association) is 0.0000 indicating virtually no effect of census on case disposition. Perhaps even more importantly, the effect that does exist occurs entirely in the high census sectors. It is not that greater percentages of patients are being admitted to fill up the hospital when the census is low, but that when the census gets too high it is impossible to admit them.

Only in Table 5, which shows the relationship between insurance and disposition from the emergency room, did we find anything like the relationship we expected, with a tendency for decisions on the basis of the hospital's best interests. Although the effect is clear and significant, it is worth pointing out that it only involves transferring to other treatment facilities 6 patients who would otherwise have been admitted.

Moreover, a close look at the text of the cases makes this look even less convincing. Of the 12 cases that were transferred or referred elsewhere, 7 were transferred to their catchment area hospital on the theory that the county would then be able to pay for their care. Two were referred to their college counseling center for outpatient treatment which would presumably be paid for by the college. One was seen as marginally in need of treatment and sent to the technical

Table 5: Disposition by Insurance Status

Insured	Inpatient	Transferred or Referred	Outpatient	Row Total
No	6	12	4	22
				5.5
Yes	210	87	84	381
				94.5
Column Total	216	99	88	403
	53.6	24.6	21.8	100.0

Chi-Square	D.F.	Significance
11.66504	2	0.0029

Lambda: (Asymmetric with Disposition Dependent) = 0.0321

training center where he was a student with the promise from the staff there that they would see that he got treatment.

This leaves only two cases in which we have some evidence that perhaps the patient's treatment did not follow his or her best interests because of insurance issues. The first case involved a manic-depressive patient from another state. Her husband called her psychiatrist at home who suggested that she get her Lithium level checked. After a thorough evaluation and a discussion with the husband the psychiatrist was debating whether or not she needed to be admitted for the night in order to check her Lithium level in the morning. In talking to the husband (HSB) the Attending Psychiatrist (A) said:

HSB: Do you plan on sending her home?
A: I'm hoping to.
HSB: Do you think she will have to be hospitalized?
A: That is not certain.... Would you have any trouble taking care of her tonight?
HSB: Maybe if you give her something to slow down it would be easier.

The Attending then went to talk to the charge nurse:

A: Well, what do you want to do with her?
CN: Well, she has no [insurance] coverage.
A: Well, that cuts down all of the options.... Since she has no coverage she can come back tomorrow. That is too bad ... I didn't think she would want to stay in the geri[atric] ward [the only available beds] anyhow.... She can come back at eight o'clock tomorrow ... if she takes two tonight [referring to the Lithium] ... can somebody see her tomorrow morning?
CN: No problem with that.

A: I don't like to conduct [a] business but....
CN: Sometimes you just have to.

There seems little doubt that the insurance status affected the decision in this case but probably only because the attending was already uncertain about what to do. Indeed, when the patient's test results came back two days later suggesting a need for medication changes, the staff offered the patient admission to the hospital in spite of the fact that the patient still had no insurance.

The second case is somewhat more difficult to analyze. The patient was an alcoholic who appeared in the emergency room late at night quite drunk. After unsuccessfully trying to refer the patient to two other programs, the clinician finally put him in a cab and sent him home. The problem is that it is difficult to tell whether the refusal to treat the patient was due to the hospital's often violated policy that it does not provide treatment for alcoholism, to the clinician's sense that the patient was not seriously interested in treatment or to the patient's insurance status. Looking at these 12 cases it seems that insurance is a substantial consideration but that the staff repeatedly sought to assure themselves that the patient's problems would be properly managed. In this context it is important to remember that 6 patients were admitted with no insurance at all.

How do we account for these statistical patterns? In an effort to gather a clearer picture of the decision-making process we selected all of the cases that have any discussion of census, insurance or catchment area issues contained in the text of the cases. While 217 such cases exist, most of them are not very interesting from our point of view, since the entire discussion consists of interactions like "Is he in our catchment?" "Yes." However, within those cases in which these issues play a significant role in the decision-making process, some clear patterns are evident.

First, each type of trouble had its standard resolution. If there were no adult or adolescent beds and the patient needed hospitalization, they tried to send the patient temporarily to a geriatric floor or schedule him or her for later admission. If the patient had no private insurance, perhaps they were in the right catchment area and could be covered by the county. If they were in the wrong catchment area, they should be transferred to their own catchment area, etc. These routines seemed to be acceptable to the clinicians.[3] They were minor deviations from the ideal, practical compromises that never evoked the comments of concern and discomfort that difficult cases such as the one discussed above in which the attending says "I don't like to conduct [a] business but...."

Just as certainly there were circumstances under which these particular routines were not acceptable. Violent or disruptive patients should not go to geriatric floors since they would compromise care of the other patients. Patients from other catchment areas who had previously been treated in the other facility and were now resisting returning there strongly enough to raise questions about whether or not they would accept treatment there should not be sent there. Likewise patients who were homicidal, suicidal or might not return could not be scheduled for a later admission date.

The following case reflects an instance in which the patient was reluctant to be treated in the hospital which her HMO wanted to treat her:

Discussion in Nurses Station:

Nurse Clinician: [HMO] won't pay anything because we're too expensive and they're going to want us to send her to F [other hospital].
Resident: Okay. Listen, what did she jump out of?
C: A second story window but some friends pulled up a truck up under the window so she didn't kill herself, she just broke both legs. She's got a history of suicide, she doesn't keep her outpatient appointments, and there's steady alcohol abuse. She says she had three good sized drinks today. I asked her to show me about how much, and I guess she's had about a pint. Her mother didn't tell me much, just said that she felt sick and that she was coming up to the hospital.
C: We have beds. One overflow on 8; two overflow on 7. Last time we admitted an overflow on 8, we transferred them to 10 and we might have to do that same thing again.
R: Yeah.
C: I see this as being pretty much alcohol related. She does say she'll jump higher next time.
R: Yeah, not much choice I guess [i.e. they can't let her go].
C: Yeah, intoxicated or not, she could do herself in. What she needs is a long term alcohol program. You know, she said, 4 years ago, that she was thinking she had nothing to live for so she cut herself with a knife. It looks like she's cut herself a lot of times....
C: We could transfer her to F, but I'm not sure they'll take her. They might not think she's really suicidal and I think if they let her go, she might jump.
R: Yeah.
C: We could put her on the Geri[atric] floor.
R: The easiest thing to do would be to have her come in here....
R: You know she wants to leave.
C: We'll have to commit her, then.
R: Can we commit her then transfer her to F?
C: If we do, the county won't transport her. It'd just be easier for us to admit her here. She was here 10 days the last time and wasn't transferred, so I figure she must be covered. I think it would be the best thing for her to come in here. The thing is, the one free bed on 10 was used by the AIDS patient, the one that went out this evening, and it hasn't been disinfected, yet. You know, maybe they could put her up in seclusion.
OBS: (The clinician phones the 10th floor, speaks briefly).
C: The 10th floor can put her on a mat in seclusion until the other room's been cleaned. It makes sense to me.
OBS: (Clinician receives a call from HMO, speaks briefly).
C: You know, F has a bed, they could take her tonight....
C: Well, how do you want to go with it?
R: Let's see how she feels.
OBS: (Staff then go see the patient in the interview room).

C: Thanks for waiting patiently. Listen, [HMO] people are probably going to want you to go to F. We could have you sent over there, if you would agree to that. Okay? Or we have no beds here but we could have you stay on a mat in the seclusion room until tomorrow. How would you feel about that?
PT: What kind of room?
C: A seclusion room. It's a room where patients go when they're upset.
PT: Okay. What floor?
C: 10.
PT: I'm not going to F. Can I go home?
C: No, we're going to need you to stay here tonight.
PT: Where?
C: On the 10th floor.
PT: Okay.

Indeed, we repeatedly find that the clinicians, in finding a management solution for the patients whom they see, are willing to compromise almost everything except the principle that patients who need immediate treatment should get it in some way. What follows is a staff discussion from another case.

Resident: He's pretty loose. Talking about razor blades and rolling in front of cars. I don't want to let him go. Based on those things I think he needs to be somewhere.
Clinician: It's no big deal to me. Anywhere you want him to go is okay.
Resident: Well, it's better to err on the side of caution. Do we have any beds?

Only when there is substantial ambiguity about the need for treatment do the variety of different financial and practical concerns begin to structure the decision-making. In the case that follows the attending physician struggles valiantly with a variety of difficult factors including his own doubts that the patient has anything wrong that psychiatric treatment might help:

Clinician: Mr. M is a 28 year old, single, male. He hasn't been in for 4 years. His problem seems to have started about a year and a half ago after his sister died. Yesterday his parents tried to commit him to E Hospital for hitting his mother. E disallowed the commitment, they said he was just drunk. His parents are upset about that. He does appear to be drunk now, his parents are here, and will say that he should be committed because of suicide. He doesn't want to be in the hospital. He says that it takes him two hours to fall asleep at night, then he doesn't get up until at least noon the next day.
Attending: Did you notice anything unusual about him mentally that we could commit him on?
C: He could do 7's ["serial 7's" is a cognitive test consisting of subtracting 7 from 100 and 7 from that result and repeating that procedure until 0] and he is oriented x3 [i.e. the patient knows his name, where he is and what the date is].
A: Any unusual behavior?
C: Not that I saw. He does report having odd body movements.

A: So we have an unusual case here.

C: It gets worse, he's out of catchment, parents are afraid of him coming home.

A: So we would have to claim depression, do we have beds? Any clue as to who will pay for this?

C: (walking over to census board) Yes, two beds on 10.

A: What do you think?

C: I think he should be in a hospital.

A: There is another answer, if his parents are afraid, we could tell them just to kick him out of the house.

C: He did say he could stay with a friend.

A: If the parents wanted to take out an emergency commitment we could stay neutral. (By this last comment the Attending was referring to the fact that if the parents don't want the patient at home they could kick him out. He already admitted that he could stay with a friend. And this would seem to get the hospital out of the problem of worrying about a case out of catchment with a dubious basis for needing treatment)

C: There's one bed on 10, two overflow beds on [a geriatric floor].

A: I see, we are going into the weekend. (The Attending was commenting on the fact that Friday is the start of a weekend and the beds might be needed for more severe cases.)

C: His parents seem to be afraid that if we send him to J [his catchment hospital] they might disallow the emergency commitment.

A: It all depends on insurance.

C: He is on Medicaid. (The Attending then tries to persuade the patient and the family that the patient will leave the family alone and live with someone else. However in the conflictual interaction between father and son around this issue, the son tries to hit the father and is committed to the hospital.)

DISCUSSION

The above evidence should not be taken to mean that we found staff of the hospital in question to be selfless angels or that the care provided was always either objectively the right thing to do or delivered with the kindness and concern of Marcus Welby. Indeed there are many examples in our field notes of disparaging and even nasty remarks about particular patients or groups of patients. There are several instances in which the staff engage in arguments with the staff of other hospitals over which of them is going to get stuck with a patient whom they did not want to treat. In one instance the clinician and attending physically prevented an ambulance driver from dropping off a patient who was sent from another hospital. In short, in doing their jobs, this group of mental health professionals show many of the same traits as other human beings doing their jobs.

The contention is not that the staff of this ER loved their patients. Rather we argue that the staff participate in what Emile Durkheim (1933) called an "occupational morality" which justifies the existence of the professions by the need to provide proper care to patients. The staff seem to be better at resisting pressures to base their decisions on financial pressures than either we or they believed. Indeed, as we noted above, in informal discussions with staff we have heard repeatedly expressed the concern that finances drive decisions. On the basis of our data in this one hospital, however, it seems that perhaps we should not uncritically accept these complaints as reflecting reality. Instead we might treat them as data. What does it mean that health care personnel seem to overestimate the power of pressures to consider the financial consequences of their clinical decisions?

On a more theoretical level, while these arguments will not be settled on the basis of one limited set of data, it seems to us that our data is much more consistent with the Parsonian position than with Freidson's. Indeed it seems to us that the staff's repeatedly expressed concern about financial pressures reflects the depth of the socialized commitment to providing "appropriate" care to their patients, rather than being an accurate reflection of practice. Repeatedly in our notes we find staff discussing the financial aspects of a case but doing little to respond to it except transferring patients to hospitals which, because of catchment area considerations, can be reimbursed for the patient's care.

In ethical terms it seems that the clinicians were treating beneficence as the dominant ethical framework in which to view their work. This is not to say that this necessarily led to a beneficent result, only that they seemed to treat the financial and institutional concerns, in most cases, as matters to be managed in order to facilitate the patient getting "proper treatment."

Before closing we need to express a final caution about the interpretation of the data. This data does not prove conclusively that clinical staff are impervious to pressures to consider financial matters in clinical decisions. Indeed, it may be that psychiatry is different than other medical areas, or it may be that the "emergency" nature of the decisions we studied insulated the decisions from administrative control. Clinicians may be more responsive to financial considerations in dealing with high cost procedures such as intensive care or transplantation. Finally, of course, this is simply one hospital. Nonetheless, we believe the data suggest that the professionally inculcated ethics to put the patient's "best interests" first will significantly interfere with efforts to reduce health costs through administrative pressures.

NOTES

1. DRGs (diagnostic related groups) form a system of medical reimbursement now used by Medicare and some other insurance systems for reimbursing medical hospital expenses and are based on the typical resource utilization for groupings of illnesses based loosely on diagnoses. Such diagnostically based groupings do not seem to account well for resource usage in mental health care.
2. HMOs or Health Maintenance Organizations are prospective payment insurance systems in which the patient pays a standardized fee in return for all needed services that fit into the initial contract with the organization. Thus the patient may have all medical, surgical and radiological services covered and up to 30 days of psychiatric hospitalization provided. However HMOs typically specify the eligible providers.

3. Such phenomena are familiar features of decision-making procedures in other contexts (Dowie and Elstein 1988; Sudnow 1965; Schutz and Luckmann 1973).

REFERENCES

Beauchamp, T.L. and Childress, J.F.
    1983   Principles of Biomedical Ethics. Second Edition. New York: Oxford University Press.
Beauchamp, T.L. and McCullough, L.B.
    1984   Medical Ethics: The Moral Responsibilities of Physicians. Englewood Cliffs, N.J.: Prentice-Hall.
Callahan, D.
    1988   Review Essay – Allocating Health Resources. Hastings Center Report 18, 2: 14-20.
Dowie, J. and Elstein, F. (eds.)
    1988   Professional Judgement. A Reader in Clinical Decisionmaking. Cambridge U.K.: Cambridge University Press.
Durkheim, E.
    1933   The Division of Labor in Society. Glencoe, Ill.: Free Press.
Freidson, E.
    1970   Profession of Medicine: A Study of the Sociology of Applied Knowledge. New York: Harper & Row.
Levinsky, N.G.
    1984   The Doctor's Master. The New England Journal of Medicine 311, 4: 1573-1575.
Morreim, E.H.
    1985   The MD and the DRG. Hastings Center Report 15, 3: 30-38.
Parsons, T.
    1951   The Social System. Glencoe, Ill.: Free Press.
President's Commission for the Study of Ethical Problems in Medicine and Biomedical and Behavioral Research
    1983   Securing Access to Health Care. Volume One: Report. The Ethical Implications of Differences in the Availability of Health Services. Washington, D.C.: U.S. Government Printing Office.
Schutz, A. and Luckmann, T.
    1973   The Structures of the Life-World. Evanston, Ill.: Northwestern University Press.
Shortell, S.M. and Hughes, E.F.X.
    1988   The Effects of Regulation, Competition, and Ownership on Mortality Rates Among Hospital Inpatients. The New England Journal of Medicine 318, 17: 1100-1101.
Sudnow, D.
    1965   Normal Crimes: Sociological Features of the Penal Code. Social Problems 12: 255-270.
Thurow, L.C.
    1984   Learning to say "No." The New England Journal of Medicine 311, 4: 1569-1572.
Veatch, R.M.
    1985   Case Study – The HMO Physician's Duty to Cut Costs. Hastings Center Report 15, 4: 13-15.
    1986   DRGs and the Ethical Reallocation of Resources. Hastings Center Report 16, 3: 32-40.

PART II

ETHICS IN THE PUBLIC ARENA

MARGARET LOCK AND CHRISTINA HONDE

REACHING CONSENSUS ABOUT DEATH: HEART TRANSPLANTS AND CULTURAL IDENTITY IN JAPAN

[History] no longer speaks of the changeless but, rather, of the laws of change which spare nothing. Everywhere history is seen as progress, sometimes sociopolitical progress, and continually technological progress....

John Berger
And our faces, my heart, brief as photos (1984)

The performing of certain kinds of organ transplants calls into question a universal event thought to be the most "natural" of all: death. By tinkering with dying we start to make what is usually taken as an unassailable division between culture and nature rather fuzzy at the seams. A comparative analysis of how different societies respond to this situation can encourage a reflective re-examination of what we in North America and Europe have come to accept as inevitable in contemporary medical care, namely, a gradual refinement in the technology and drugs associated with organ transplantation, and hence an increase in the number and further routinization of such operations.

The current ethical debate in connection with organ transplants includes some extraordinarily delicate topics such as the status of anencephalic infants. Nevertheless, debate in the West around such issues is grounded in the assumption that organ transplants are now "normal" and inevitable, and that it is therefore essential to maximize the availability of organs for transplantation. Much of the literature is at present focused on questions as to whether a "contract-in" or "contract-out" system is most appropriate for increasing organ donation (Somerville 1985); whether adoption of a market model for obtaining organs is appropriate or not (Prottas 1983; Williams 1985); and whether the body should be considered a form of property (Andrews 1986).

Perhaps it is necessary to move once in a while during the course of this important debate to a more distant observation point and ask ourselves why we have arrived, apparently so quickly, at the position of accepting organ transplants as routine. In order to address this basic question it is necessary to go beyond legal discussions and market models and enter the more emotionally laden world of values. We can show, for example, that so-called neutral, ethical arguments in connection with organ transplants are grounded in unexamined values which, among other things, promote routinization. One can also ask why the "official" debate has been dominated by input from "experts" and why discussions which incorporate the viewpoints and experiences of nurses, junior medical staff, organ recipients, and family members of both donors and recipients are, with few exceptions (but see Youngner et al. 1985) virtually non-existent.

In Japan, where, of course, the necessary technology for organ transplants is available, routinization of these procedures has not taken place; on the contrary,

G. Weisz (ed.), Social Science Perspectives on Medical Ethics, 99–119.
© 1990 Kluwer Academic Publishers. Printed in the Netherlands.

despite a massive national debate on the subject, brain death[1] is not generally accepted as an adequate definition of death, and transplants are limited almost entirely to those which do not require a brain-dead donor. The Japanese case therefore provides a cautionary tale about the heterogeneity of value systems in the post-modern world, and an illustrative example of how the acceptance and application of technology can only be understood in cultural context. At the same time, this study can serve the classic anthropological function of stimulating some reflective thoughts about ourselves.

With the hindsight of experience, it is now clear that explanations about modernization and social change on a global scale which assume a gradual homogenization, an irreversible blurring of difference in the bright light of reason, are over-simplified. If there is eventually to be a dominant world ideology, its unfolding will not be a simple process. Since science is not liberated from cultural construction, there are many sciences distributed throughout time and space (Lock and Gordon 1988; Needham 1962; Wright and Treacher 1982). Moreover, in recent years, Western science has come to be associated in many parts of the world with a new kind of imperialism. In an era of nationalisms and the creation and recreation of cultural identities, national and regional sciences become part of the baggage which is used to establish uniqueness, difference, and local bases of power. We cannot consider the application of technology, therefore, as the simple unfolding of scientific progress, neutral and unavailable for cultural analyses.

We would like to take up two points for discussion in this paper. First, the particular issue of defining death and its relationship to the carrying out of organ transplants is illustrative of the extent to which cultural values and social organization can severely curtail the "technological imperative" said to be associated with all post-industrial societies (Fuchs 1974). Nowhere is a dead body merely a cadaver, readily available for butchery, abandonment, or experimentation. Nor can an individual death be a purely private issue. All societies have seemingly deemed it necessary to institutionalize certain practices and rituals in connection with their dead. According to Lifton et al., such rituals "reflect widespread psychological needs for placing death within a larger context of collective human continuity" (1979: 144). These rituals are not merely about the loss sustained by family members, they serve simultaneously to reaffirm basic societal values about human life and the relationships of individuals to the social order. Emotionally laden culturally variable ideas about, for example, the boundaries of culture and nature, humanity and raw biology, the individual and society, the relationship of mind to body, and so on, form the background against which debates about death and the sanctity or otherwise of the dead body take place. We will use the current debate about brain death in Japan to illustrate this point.

Secondly, it will be shown how in the present debate about death in Japan, it is often stated that the "other" (most usually defined as America) is a place where conditions differ from Japan in many fundamental ways, where dehumanizing medical practice is frequent, and which therefore should not be taken as a model for Japanese medical knowledge, practice, and ethics in the late 20th century. This type of argument is not limited to the issue of organ transplantation, but is used with great frequency in connection with any aspect of modern life in Japan where it is believed that traditional values are being eroded by a flood of Western values, most especially that of individualism (Lock 1988).

THE NEED TO DEFINE DEATH: NORTH AMERICA AND JAPAN

The first respirators were developed during the 1950s, making it possible to sustain brain-dead but otherwise alive human bodies for anything from a number of hours to, occasionally, several months or even years. At this time too, kidney transplant techniques were developed and then, after the first heart transplant was carried out in South Africa in 1967, demand for fresh human organs spiraled; it thus became an urgent priority to clarify a concept of death for use in the modern medical world.

It has been pointed out that attempts to redefine death from the late 1960s to the mid 1970s, including a 1966 London CIBA Symposium and a World Medical Association meeting in Sydney in 1968, worked under the assumption that what was being done was unprecedented and a necessary response to new medical technology (Pernick 1988; Rado 1981). Pernick, in an illuminating article shows, however, that the recent concern about how to define death is not new, but merely the most recent reemergence of an issue which has been revived repeatedly throughout medical history. Moreover, difficulties in defining and diagnosing death have often sprung, in the past as now, "from new medical discoveries, especially in such areas as experimental physiology, resuscitation, and suspended animation" (1988: 17). Pernick cites a 1940 article in *Scientific American* which claimed that "frequent" errors in diagnosing death, a perennial issue for concern, were still responsible for cases of premature burial (Newman 1940). Pernick's article shows moreover, that it is not merely responses to specific medical discoveries which have shaped the content of debates about death. Professional interests and social values are inevitably implicated and preclude the possibility of an objective definition.

The first step taken in North America in response, on the one hand to the development of respirators and, on the other, to increasing pressures to carry out organ transplants, was by the Ad Hoc Committee of the Harvard Medical School, a group of doctors who wished to promote the practice of organ transplants. This committee declared in 1968 that individuals in a state of "irreversible coma" who were diagnosed as having "brain death syndrome" could be declared dead (Ad Hoc Committee of the Harvard Medical School 1968). During the early 1970s the concept of brain death was challenged in the courts. In one landmark case in Virginia in 1972 the jury ruled against the donor's family who claimed that the transplant surgeons had been responsible for the death of their relative. Other court cases followed including several involving homicide victims (Simmons et al. 1987). At the same time a debate about actual medical practice was carried on, most particularly in connection with which tests, if any, could be relied upon to confirm an individual doctor's opinion of brain death; whether EEGS were adequate indicators of brain activity or not, and who should be the "gatekeepers" to protect physicians from malpractice suits. The heat of this debate was felt most clearly at the level of state legislatures and in tertiary care hosptials.

As a result of the uncertainty which surrounded the entire discussion, a President's Commission was set up in 1980 which set out to define a Uniform Determination of Death Act. This Commission, in opposition to the stance taken by many physicians, philosophers, theologians, and others, opted to update so-called "obsolete" criteria (Zaner 1988: 2). It was recommended that a concept of "whole-brain death" be adopted which was distinguished carefully from a "persistant vegetative state" as exemplified by patients such as Karen Ann Quinlan whose brain stem continued to function despite irreversible loss of higher brain function. The higher brain, that most vulnerable of organs, is the first to die if the flow of oxygen to the body is cut off. Cerebral death corresponds, then, to a permanent loss of consciousness and of "all manifestations of personality" (Law Reform Commission of Canada 1979), but occasionally, if the lower brain stem remains undamaged, patients can retain a gag reflex, swallowing motions, and the urge to breathe.

In a recent publication a group of philosophers, lawyers, doctors and an historian take a critical look at the Commission's recommendation (Zaner 1988). The view of the Commission is that death can only be declared once the body has lost the ability to "organize and regulate itself" (President's Commission for the Study of Ethical Problems 1981: 32). All brain function must have ceased, including the reflexes controlled by the brain stem. At this point there is a lack of neurological integration and what is left is "no longer a functional or *organic* unity, but merely a *mechanical* complex" (Bernat et al. 1981: 391). As Zaner points out, for whole-brain advocates, "it is the biological organism (or, more specifically, the physiological/anatomical nervous system) which is definitive for life and for death, not the *person* whose organism (or nervous system) it is" (1988: 7). He and the majority of the contributors to the book *Death: Beyond Whole Brain Criteria* believe that the Commission has put the cart before the horse in that it has tried to develop a concept of death out of a set of standardized medical tests while at the same time evading the central issue of just *who* has died. In order for there to be a uniform statutory definition of human death there must be first of all, states Zaner, a general consensus over what constitutes "personhood" or "personal identity"(1988: 5). The Commission acknowledges that there is no such consensus and that the question has been debated for centuries but it attempts to side-step the issue using a reductionistic argument in which operational criteria and medical tests provide the answer. Bartlett and Youngner reply that one can establish neither operational criteria nor valid tests unless one has agreed upon a working definition; a concept of what it means to die (1988). This concept has to be societal and not biological since, they argue, it is the permanent loss of personhood which is of central concern. Such an argument leads us out of the usual legal and medical worlds into the realm of morals and values. Simultaneously, a move is made away from dichotomous variables, truth and falsehood, categories of inclusion and exclusion, to the shifting, contingent arena of contextualized decision-making. From such a vantage point Zaner and his colleagues can say that, "at the very least ... the central issues inherent to any definition of death must be kept rigorously open" (1988: 13).

Recent reports and letters which have appeared in medical journals in North America indicate that medical professionals still have very mixed feelings about defining brain death and making decisions about when to disconnect the respirator. For example, arguments have been published as to whether unusual spontaneous movements are normal after the respirator is discontinued, or if such movements indicate that the patient had been wrongly diagnosed as brain-dead (*JAMA*: 1986). The most topical debate is in connection with the need to decide whether anencephalic newborns can be defined as brain-dead despite the absence of all but the brain stem. Physicians are obviously concerned about the tremendous responsibility they bear in connection with decisions about death. For many of them, the philosophical arguments about personhood are too academic; the dramatic and graphic scenes which they witness in medical practice can drive physicians into an essentially defensive posture in which they put their faith in the establishment of universal, scientific criteria.

It has been pointed out that the decision to accept either cerebral or whole-brain death as the border line between life and death has been a painful and difficult one to reach. Brain-dead but otherwise alive human bodies are warm to the touch and are respiring, albeit by mechanical means. Stating the position often taken by opponents of brain death acceptance, the bioethicist Tristram Engelhardt has said, "[Brain-dead bodies] appear to be alive because they are in fact alive" (1986: 209). Such alive/dead persons could, in theory, act as sperm donors, and healthy babies have been delivered from females in this state (Shrader 1986). Organs removed from brain-dead patients still on respirators transplant much more successfully than those removed after heart death; in fact, hearts for transplants can only come from patients in this state.

Despite the fact that there has been considerable medical and legal debate as to whether brain death can be established beyond a doubt, this topic, which clearly touches every one of us in some way, has not been a subject of much *public* debate in North America. (This situation, however, has changed somewhat recently with the reporting in the media of the use of organs from anencephalic babies for transplants.) Some states and provinces have established legal definitions of brain death, others have not. Neither the American nor the Canadian supreme courts have passed judgement on this matter. Nevertheless, virtually all of the larger hospitals in the major cities of North America and most of Europe now carry out organ transplants on a routine basis. (As of this year over 9,400 heart transplants have been carried out worldwide.[2]) In North America a continent-wide computer network keeps hospitals informed of brain deaths (many of the donors are young victims of road accidents) and teams of experts fly harvested organs or brain-dead bodies around the country as required.

The situation in Japan stands in stark contrast to that in North America. The first and only heart transplant to be carried out in Japan to date was performed in 1968 in Sapporo. The same medical team which determined brain death also conducted the transplant, and it was agreed in the national debate which ensued that the research interests of the doctors had interfered with their care of the donor. The senior surgeon was prosecuted on a charge of murder, but was later acquitted. Since that time, there have been no heart or lung transplants and only two liver transplants.

The issue remained relatively dormant in the years following the first organ transplants, but in 1979 cyclosporin-steroid therapy was introduced into clinical medicine in both Japan and the West. This drug has proved extremely effective in reducing the rejection rate of transplanted organs and therefore provides an enormous incentive to perform this type of surgery. A sense of urgency about confronting the issue was generated in Japan. More recently, a highly controversial kidney/pancreas transplant case at Tsukuba University involving a woman suffering from "mental problems" whose permission to act as a donor had not been obtained prior to her death (*Mainichi Daily News* 1984), has inflamed the debate. Added to which some Japanese have expressed embarrassment because a few of their compatriots have been recipients in America of donated organs which are said to be in short supply. These issues have ensured that the subject of death is currently being given very high priority in the media in Japan.

THE JAPANESE DEBATE

The debate in Japan has been punctuated by several oft-repeated themes, the most prominent being that there must be public consensus before the status quo can be ended. One Japanese researcher has stated that problems such as this, which in the West are considered the province of bioethics, are dealt with in Japan as a matter of creating an "etiquette" which will be generally accepted (Yonemoto 1986).

Engelhardt has pointed out that in large-scale, pluralistic, secular states one can only hope for a "general commitment to peaceable negotiation as [a] cardinal moral canon" (1984: 67). Reaching a relatively clear public consensus is not something that can be expected given the plethora of religious, moral and ethical heritages which constitute modern North America. Japan, in contrast, can resort to time-honored methods for producing social change, methods to which an educated public and principles of consensus are central. In a society where only 1% of the population is of foreign origin, where education is standardized and centralized, and the religious and philosophical heritage is shared in common, the principle of consensus can, in theory at least, be a realistic ideal.

Every discussion on the subject of brain death in Japan, even when presented by doctors eager to facilitate organ transplants, is based on the premise that any changes in medical or legal practice in this area can only come after a consensus has been established among the Japanese people as a whole. To this end there has been regular media coverage of the subject and several national opinion polls have been taken. One published by the Prime Minister's Office in the fall of 1987 showed 61% of the respondents "interested in the question of brain death," but only 23.7% unconditionally accepting brain death as death of the individual; 36.7% said it should be up to the previously expressed wishes of the patient and the judgement of his family whether to accept brain death, and 24.1% maintained that death should not be pronounced until the heart stops (*Asahi Shimbun* 1987a).

It has been claimed that arguments in favour of the acceptance of brain death as the end of individual life have been given more media publicity than those against it, and that the Japanese public is being manoeuvered into believing that most people are willing to accept the concept (Nakajima 1985: 259). Over the past three years, however, the opponents' point of view has also been widely aired. Articles and books on the subject cover in detail the purely scientific controversies as well as the cultural and social arguments pro and con. For example, a recent article in the Asahi newspaper (which has the largest circulation in the world) about a conference of the Japan Association of Brain Surgeons, reported on a research paper which posed questions about the standards for brain death recently established by the Ministry of Heath and Welfare in Japan. The article elaborated *in minutiae* about hormone circulation and reflexes after brain death, and was illustrated with photographs of spinal cells from brain-dead patients (*Asahi Shimbun* 1987c). This type of evidence (which is given very little attention by the North American media) is used to alert the public to the controversial nature of the scientific evidence being drawn upon to justify acceptance of brain death. It is frequently suggested in articles such as these that vital decisions with major implications for medical practice should not be left up to the medical profession alone.

Numerous legal and scientific bodies have considered the question to be one about which they should elaborate a position. In October 1987, the Science Council of Japan, an advisory body to the Prime Minister, discussed the issue of brain death: the medical section recommended strongly the acceptance of brain death, but opposition from the other disciplines represented on the Council prevented any final unanimous decision (*Asahi Shimbun* 1987b). The Japan Society for Transplantation and the Life Ethics Study Parliamentarians League have declared themselves in favour of accepting brain death, while the Japanese Society of Psychiatry and Neurology and the Japan Federation of Bar Associations oppose its recognition.

In January 1988 the Life Ethics Council of the Japan Medical Association recommended that whole-brain death be accepted together with heart death as equivalent to individual death. A group of doctors at Osaka University immediately applied to the ethics council of the university for permission to carry out heart transplants, and doctors at twenty other hospitals have followed suit. This was followed rapidly by a national survey of 3000 informants conducted by the Asahi newspaper. The results of this survey which had a response rate of 78% were reported in a two-page spread on the subject. The headline states that there

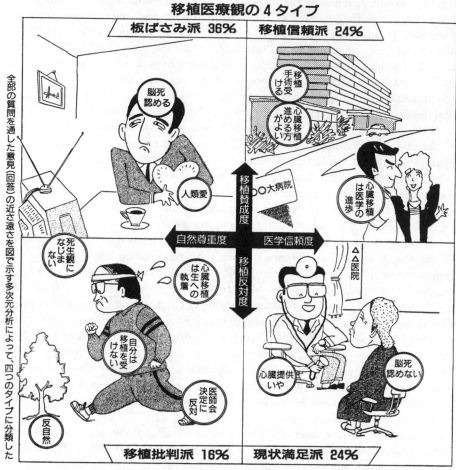

移植医療観の４タイプ

板ばさみ派 36%　移植信頼派 24%

移植批判派 16%　現状満足派 24%

is an increase in the number of people supporting the institutionalization of heart transplants (*Asahi Shimbun* 1988a). However, the comments of Kazuo Takeuchi, chairman of the research group appointed by the Welfare Ministry to discuss brain death, are published together with the results of the survey. Dr. Takeuchi states that it is hard to interpret the results, and goes on to point out the many discrepancies in the way in which the questions were answered (*Asahi Shimbun* 1988b; see above and translation p. 107). Writing in a widely circulated weekly magazine, Ichiro Katō who was until recently the chairman of the above-mentioned Life Ethics Committee of the Japan Medical Association states that the idea of consensus is a "mirage." He believes that the concept of national consensus, prevalent throughout traditional East Asia, is just a "sneaky" way to avoid difficult issues and to put off making decisions. He urges people to start thinking more individually in Japan, to be less manipulated by the media, which,

Four Types of Observations in Connection with Transplantation
(Translation of Figure on p. 106)

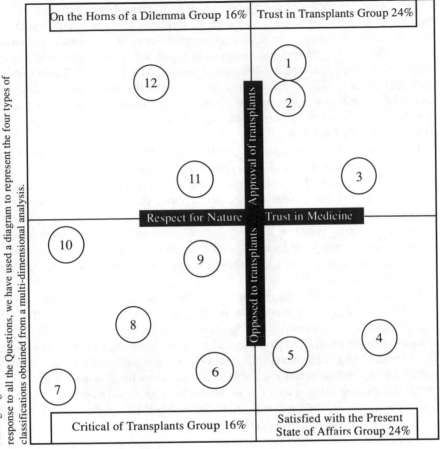

1 - It's alright to have a transplant operation.

2 - Many people want progress in heart transplants.

3 - Heart transplants will help medical progress.

4 - Doesn't recognize brain death.

5 - Doesn't want to donate his heart.

6 - Opposed to the decision of the Japan Medical Association.

7 - It's against nature.

8 - I don't want to receive a transplant.

9 - Having a heart transplant means hanging on to life.

10 - Doesn't fit with Japanese ideas about life and death.

11 - For the love of Humanity.

12 - Recognize brain death.

he states, is largely responsible for promoting the need for public consensus in connection with brain death (Katō 1988).

Thus, both supporters of the promotion of heart transplants, such as Katō, and opponents, such as Nakajima, claim that the public is being manipulated, largely by the media. Supporters believe that negotiations for consensus are simply slowing the advance of medical progress in Japan and at the same time serve to promote an antiquarian image of Japanese medicine to the world. Those opposed state that there is little evidence from the surveys that the public is indeed in favour of heart transplants, nor that opinions have changed much over the past few years despite all the publicity. Nevertheless, the media claim that gradually consensus is being reached.

In theory, at least, the Japanese public should by now be highly conversant with the scientific and ethical issues surrounding the definition of brain death. However, since most of their information is obtained in a rather haphazard and sometimes biased form from newspaper reports, this is doubtful. One of the surveys carried out by the Asahi indicates that the public is confused, since 38% of the respondents stated in answer to one question that they would want to receive a heart transplant should they themselves need it, but in response to another question, they stated that they are opposed to the acceptance of brain death (*Asahi Shimbun* 1988b).

Despite the numerous applications to carry out heart transplants, and the established legality of the procedure, in the year since the acceptance of whole-brain death by the Japan Medical Association only two transplants have been conducted using brain-dead donors, and both of these were kidney transplants. The hospitals which performed the procedures received extensive media exposure. Clearly, certain inhibiting factors still seem to be at work. Amid all the public debate, one thing stands out in the majority of articles and commentaries: while it is agreed by the Japanese that their society is culturally distinct from the West, the actual debate about death is largely focused on the same issues which dominate the North American literature, namely, the establishment of clear, scientifically valid criteria and tests. Indeed, even more emphasis appears to be given in Japan than in the West to the question of residual reflexes and so on. With the exception of a few writers (to be discussed below) obvious cultural differences are not very often discussed. However, the very way in which the so-called scientific data are interpreted in Japan, leads us right into the question of the relationships among culture, scientific knowledge and the application of technology.

THE CULTURAL CONSTRUCTION OF DEATH: ANCESTORS AND IMMORTALITY

Although the debate is generally carried out as though it is resolvable through the application of basic science and statistical methods, there are numerous cultural and social factors which mitigate against an easy acceptance of brain death and organ transplants and which have a profound impact on public discussion in

Japan. Commentators are not agreed among themselves as to which factors, some of which are emotionally laden and difficult for them to articulate clearly, are the most relevant.

It seems highly likely that in cultures which have thoroughly embraced the Cartesian mind/body split, and where "human-ness" is closely associated with the mind, often to the neglect of the body, that it will be easier to accept brain death as final. The report of the Law Reform Commission of Canada on the subject states, "There now seems to exist a recognition of the idea that death is the termination of individual and relational life, and that therefore there is no point in prolonging a merely biological existence once personal life, as such, has been irretrievably lost" (Law Reform Commission of Canada 1979: 5).

In contrast, the Japanese anthropologist Emiko Namihira has commented, "Japanese view the body and spirit as mutually connected" (*Igakkai Shimbun* 1987), and the word *kokoro*, meaning "spirit," is written with the character for "heart." In this context, it would make sense that an individual cannot be declared dead until that "core" of humanity, the *kokoro*, has ceased to function.

The idea of "personhood" is something which has been considered not only by philosophers, but also by anthropologists who have shown that a relativistic understanding of concepts such as individuality, self, and personhood is necessary if we are to develop a comparative account of human behaviour (Levy 1973; Rosaldo 1984). Human beings are both biological organisms and also persons, and, comparatively speaking, the characteristics which are identified with personhood may not be valued in the same way as those identified with the physical body. Shrader has proposed that "we regard a *human being* ... as a composite of two intimately related but conceptually distinguishable components"(1986: 13). He suggests, therefore, that we are subject to more than one death, that of the person and that of the biological organism. In the post-Enlightenment West, the biological organism has come to be considered as best understood through the application of science; as such it is regarded in a relatively mechanistic fashion as a universal object and independent of time and space. The "person," in contrast, represents the essence of individuality, of uniqueness, and of humanity (Wartofsky 1988: 219; Young 1988) and is for most people (I assume) located mainly in the brain.

In Japan, the biological organism is much more individually conceived, with emphasis given to the unique constellation of physical and psychological predispositions with which each human is believed to be endowed at birth. The Japanese "person," on the other hand, is a finely honed social being who is trained from birth onwards in conformity, dependency, and the suppression of individuality (including any biological weaknesses) for the sake of group harmony (Lebra 1976; Lock 1980). The private, "unique," and most humane part of a Japanese is not equated with the "person," but is usually associated with the highly evocative *kokoro*, located somewhere in the chest or thorax. It is not surprising, therefore, that the acceptance of brain death is a very controversial issue in Japan.

Although not all informants[3] were in agreement on this point, it is generally believed that the fate of the body after death and the anthropomorphization of ancestors contributes to a reluctance to both donate and receive organs. Dying is visualized as a process in Japan and takes place in stages. According to Buddhist beliefs, for the first forty-nine days after death the spirit of the deceased remains in this world. It would be highly inappropriate and very troubling to many families to know that they had allowed parts of the body of a dead relative to be removed. The fact that many Japanese volunteer to donate only one cornea at eye banks can perhaps be interpreted in light of the ancestor cult (Feldman 1985).[4]

Many people point out that although they do not actually believe in Buddhism, family and social obligations require that deceased family members be treated with respect in accordance with Buddhist ritual. It takes a total of thirty-two years for a deceased family member to fully merge with the community of spirits and become an ancestor, and there is an ongoing cycle of rituals to be acted out by descendants in order for the process to be successful. During this time and usually for many years to follow, a photograph of the dead relative is placed in the *butsudan* (the family altar kept in the majority of households headed by the eldest son). The ancestors are the guardians of moral order and as such are accorded due respect: family members talk regularly (often literally) to them, and also leave ritual offerings of food for them. The image of the deceased is therefore kept very much in the present, and visualized in human form. Although physical death may have taken place years previously, one attains social immortality as an ancestor.

Namihira has pointed to a subtle but important language usage in connection with a dead body. The Japanese word *shitai* refers to the corpse, but the word *itai* (with an honorific in front of it) is used in preference to *shitai* when one is talking to family members about the deceased, or whenever the relationship of the body to living relatives is discussed (1988: 44). The concept of *itai* includes the ideas that one should retain feelings of attachment to the dead person and also, theoretically, that the dead person in the form of the *itai* can make demands on living relatives (Namihira 1988: 46).

Namihira analysed interviews which were carried out with relatives of victims of the Japan Air Lines crash in the Gumma mountains in 1985 and draws some conclusions about contemporary Japanese views about death from the very moving narratives (1988). She concludes that the spirit of the deceased is anthropomorphized, for example, and is believed to experience the same feelings as do living people. Living relatives have an obligation to make the spirit happy and comfortable. There is a consensus that a dead person wishes to be brought home, and that a corpse should be complete (*gotai manzoku*) otherwise the spirit will suffer and is potentially able to cause bad luck for the living. Namihira cites a 1983 questionnaire carried out by a committee set up to encourage the donation of bodies for medical research. Out of 690 respondents 66% stated that cutting into dead bodies is repulsive and/or cruel and/or shows a lack of respect for the dead. In this same study 66% of 685 respondents stated that there are no religious beliefs in their area which influence ideas about the dead, indicating, presumably, that most people think it "natural" to believe in spirits. Other observations which emerged from this study are: a concern with what other people will think if one allows the body of a relative to be used for medical purposes; and an interest in having hospitals pay for the very expensive funeral costs if one co-operates in organ donation.

Not only Confucian ethics (from which the ancestor cult probably springs) but also the basic values of Shinto (the indigenous animistic Japanese religion) are opposed to tampering with the human body. Confucian-derived ideas apparently account in part for a reluctance to donate organs. Shinto-derived beliefs, on the other hand, are cited by Japanese informants as having an influence on the harvesting of organs for transplants. Handling of the dead is associated in the minds of many with concepts of ritual pollution (*kegare*). Historically, dissection of bodies was forbidden in Japan, as in the West, and the earliest dissections were carried out on dead criminals in the 17th century by Dutch doctors residing in Japan. When discussing the subject of organ transplants, many informants are clearly visibly repulsed and state that the whole idea is "unnatural." Putting money into the development of artificial organs is considered to be much more appropriate, and has attracted the attention of many Japanese researchers.

Traditional ideas about individual versus family rights also inhibit the donation of body parts in Japan. The body of a dead person is believed to belong as much to the family as to the individual and therefore the signing away of parts of one's body for medical purposes without first consulting the family is not usually acceptable. Intimately involved with the question of family obligations and rights over the human body are the ever-present constraining forces in Japanese society in connection with giving and receiving. Yonezo Nakagawa, professor in the medical faculty of Osaka University, has commented, "Brain death and organ transplants are technology which originated in the West, and are not acceptable in Japan ... Organ transplants are said to originate in love of humanity, but the basic idea is that of taking something that belongs to someone else, and I doubt that many Japanese will approve of passing a law to allow it" (*Yomiuri Shimbun* Editorial Department 1985: 83).

The lack of a tradition of altruistic giving to unknown others, coupled with the force of Confucian derived obligations within one's family and one's group, have created a uniquely Japanese situation concerning kidney transplants: while the majority of kidney transplants in the U. S. and Europe come from cadaver donors (69% in the U.S., 91% in Europe), in Japan the majority of kidneys transplanted (74%) are donated by living relatives (Katō et al. 1986: 12). In addition, at least one Japanese opponent of liberalized rules for organ donations fears that people may be unwillingly pressured into donating the organs of a family member by their feelings of obligation to the doctors who cared for the patient (Nakajima 1985: 218). Starzl has made a somewhat similar point in connection with the West. He has suggested that in a Judeo-Christian society in which self-sacrifice is a working ethic, more courage may be required to resist organ donation between family members than to accept it (1985: 5). Both Japanese and Western family members may be "at risk," therefore, although the motivating forces come from radically different philosophical and religious traditions.

In the West, a tradition of a right to individual choice, coupled with a Christian tradition of altruistic giving, has probably facilitated the ready acceptance of organ donation, and in recent years our drivers' licenses are testimony to how widespread this custom has become. In Japan, drives to encourage the donation of organs and even of blood have had little success, and commentators are sensitive to the fact that there is no long tradition of altruism on which to draw. Namihira is quick to point out that there is variation in Japan and that not everyone is reluctant to donate organs (1988: 63), nevertheless, in comparison with North America, there does seem to be considerable resistance.

THE JAPANESE HOSPITAL AND MEDICAL PROFESSION – THE CONSTRAINTS OF CONVENTION

It has been pointed out by a Japanese legal expert who has written a popular and well-informed book on the subject of brain death, that there is financial incentive in the U.S. to remove relatives from the respirator since families must pay for hospital care (Nakajima 1985). In Japan, although the State pays for hospital care, there would appear to be another kind of incentive to hasten the demise of a cerebrally dead relative; that is, family members must spend many hours doing much of the necessary nursing even inside the hospital, and it is not at all rare to find an exhausted wife bedridden in her turn after having nursed her husband through his last illness. However, the very fact that the Japanese family, unlike the American family, is intimately involved with the dying person is probably incentive enough to try to sustain life against all indications to the contrary. The medically "informed" North American family is usually kept at a distance and hence somewhat ignorant of what actually takes place in many hospital settings.

While brain death may appear to be definitive in a scientific sense, it is extremely difficult for the layman to accept the idea that the warm, breathing body of a relative is medically "dead." Because relatives play a major role in patient care in Japanese hospitals, they are more likely to be present in large numbers at the time of death than are family members in the West, and the question of the family's emotional acceptance looms large in many Japanese doctors' discussions of the subject. They claim that they are reluctant to approach families to ask for permission to remove organs. It is customary in Japan for the family to wash and lay out the dead relative, even in tertiary care hospitals, a custom which would clearly be severely curtailed by the arrival of a team of organ "harvesters." The body is then removed to the family home for commencement of Buddhist ceremonies. An absent body, flown to the other end of the country as a donor would be unthinkable for most families.

Constraints against the widespread practice of organ transplants arise not only from patient families but also from within the medical profession. Japanese surgeons are internationally acknowledged to be extremely skillful. Moreover, transplant surgery could contribute substantially to the profits of their hospitals (most of which are private) as well as to the prestige of the doctors. Nevertheless, there appears to be considerable reluctance within the medical profession to carry out transplants. It is striking that even though brain death has been legally acceptable now for over a year, it has so far made virtually no difference to medical practice.

It is frequently reiterated that a Japanese doctor should strive to prolong life at all costs, and there are many examples of the "heroic" measures taken by physicians to keep patients "alive" to the bitter end (Long and Long 1982). It is extremely difficult for many Japanese doctors to admit that they ever unplug a respirator, even though a study of physician attitudes in Boston and Tokyo showed that 31% of doctors in both countries support the legalization of euthanasia. However, in actual practice it was shown that when Boston doctors administer pain relievers which are known to accelerate the process of dying, only 48% of them continue life-support treatment at the same time, whereas 89% of the Japanese doctors stated that they did so (Fukuda 1975). There seems to be considerable ambivalence, then, among Japanese physicians; current literature on the subject indicates that the profession is rather bitterly divided on the question of accepting brain death as the end of life, especially if that would mean obligatory removal of respirators (*Mainichi Shimbun* 1988). Even those who are in favour of change believe that they should not take such an enormous responsibility onto their own shoulders, and they are therefore willing to bide their time, even at the expense of the advancement of their own careers.

One final, but perhaps crucial factor which may have speeded up the North American side of the story is that, compared with Japan, there are many more road accidents and gunshot deaths involving people under 25, that is, those who are the most suitable organ donors. A rather macabre, but not altogether surprising point of view in some circles in North America, is that the harvesting of organs from young "healthy" but unfortunately "dead" bodies, makes sound economic sense since it clearly makes a contribution to scientific progress.

NATIONALISM AND THE EVOLUTION OF BIOETHICS

One point which comes up quite frequently in the Japanese debate is that the bioethics developed in the United States should not be assumed to be appropriate for Japan, that the latter nation has long since relinquished the role of imitator and is now forging ahead in the creation of its own modern state and cultural identity. It is explicitly stated that simply because heart transplants are routinely performed in the United States does not mean that they should *ipso facto* be carried out in Japan. Keichi Amano claims, "There is a great difference between the cultural ethos of the West and that of Japan.... Although it is easy to copy the material aspects of the other countries' cultures, as far as the ethos is concerned, foreign is foreign ..."; he goes on to suggest that, rather than copy the West in the field of transplants, Japan should use its technological prowess to produce an efficient artificial heart (Amano 1987: 1949).

In the course of research for the present paper, we came across statements in the Japanese literature which allude to unethical and dangerous practices in the

United States in connection with organ donation. Amano, writing in a respectable nursing journal, states flatly that American families who donate organs of brain-dead relatives receive enormous sums of money, a contention absolutely denied by every North American doctor we have questioned. The article in question claims that in America, "where money is everything," a poor black woman in Harlem would think nothing of donating the organs of a child of hers killed in a road accident, implying that her motive would be the compensation she would receive (Amano 1987). A second example appears in a book entitled *Brain Death and Organ Transplants* (*Yomiurui Shimbun* Editorial Department 1985: 42), in which the story is told of a false diagnosis of brain death such that the patient was still breathing independently when the transplant surgeon arrived at a St. Louis hospital to rip out his heart. It is interesting to speculate as to whether there can be any grounds for these stories.[5] What is of even more interest is that there has been no mention of them in the American media and that they are being cited in Japan by transplant opponents who have taken it upon themselves to hold up the United States as the evil "other" to their compatriots. What passes for "ethical" in America is not good enough for modern Japan, a country where the poor and ethnic minorities cannot be exploited in such a way! According to one commentator:

> It seems that many things happen [in America] which would be impossible in a country like Japan where the standard of living and educational level are basically the same all over and the entire country reads the same news and the same editorials in national newspapers. It is fine to adopt the good points of other countries, but it seems to be about time for Japan to escape from the idea that it will be a backward country if it doesn't copy the West ... (Nakajima 1985: 258).

Ichiro Katō states that individual freedom is a fundamental principle in the West, whereas in Japan it is the idea of regulation (*kisei*) which predominates. Katō believes that in the West one is free to carry out organ transplants, for example, unless specially prohibited by law, whereas in Japan one cannot undertake such procedures without official approval. He is impatient with the inevitable delay which sets in if one waits for consensus to be reached about every major new technological development. Nevertheless he is in agreement with the dominant feeling that the West can no longer serve as a model for social change in Japan (Katō 1988).

LESSONS FOR THE WEST

It is now legally possible to do transplants using brain-dead donors in Japan, but the debate for and against continues to make headlines and no heart or liver transplants are being performed. Meantime more than fifteen years of medical experience have been lost (although several physicians have received training in North America). At the end of this prolonged cultural debate there will be, in contrast to North America, a public which is sensitive to the issue, many of whose members will believe (perhaps incorrectly) that they have participated in creating the new policy using the characteristic Japanese means of working towards consensus. The result will probably be that there will be few legal battles or disputes around this issue in the future.

In addition, it appears that acceptance of brain death as definitive will continue to need the consent of the patient's family, and that organs will not be harvested without prior consent of the donor (wherever possible) and the family. This may help to promote the concept of informed consent on a general basis in medical practice in Japan, something which so far has not been the case. It may well also account for the fact that there have been so few transplant cases involving brain-dead donors despite the legalization of the procedure.

The urge to participate in this most modern and dramatic form of medical technology is clearly very strong among Japanese surgeons. No doubt they would be extremely innovative and successful given the chance. For the foreseeable future, however, they must on the whole content themselves with experimentation with artificial organs. It seems likely to us that while a careful separation is maintained between those physicians who care for dying patients, and the transplant team, the cultural constraints against obtaining permission from families, and the revulsion of the families themselves will act against any rapid increase in the use of brain-dead donors.

What can we learn from the Japanese experience? The most obvious lesson is perhaps that given the pluralism of North American society, one would not expect a ready acceptance of brain death among all physicians or all members of the public. An ongoing public debate in the media on this side of the Pacific involving public personalities from all walks of life (as is customary in Japan) in addition to professionally interested participants might prove highly informative. Much more publicity concerning the extent to which transplants are carried out, who gives and who receives organs and under what circumstances, and what the survival rates are, is information that everyone should be reasonably well acquainted with.

We would also like to see the experiences of nurses, junior doctors, next-of-kin, and organ recipients being made more public, something which is not done anywhere on a regular basis to the best of our knowledge. What is the psychological effect of close involvement with brain-death decisions? It has been reported that nurses left alone with an "empty shell that was once a person [after the harvesters have finished with the body] ... may feel resentment or envy of staff members who can 'get out when they can'" (Youngner et al. 1985). Moreover, it is often junior staff members and nurses who are the most involved with obtaining consent and consoling family members; little research has been undertaken into their experiences with and attitudes toward this difficult role.

It has also been pointed out that no research has been done on the effect on the grieving process when final partings are made in such unusual circumstances (Youngner et al. 1985). Similarly, the long term effects on donors and receivers of organs still await investigation. Studies using an ethnographic approach involving the narratives of the various participants (along the lines of the early study carried out by Fox and Swazey 1973) would provide invaluable information in trying to create better guidelines for dealing with the ambivalences, aversions, and psychological repercussions that are inevitable as we proceed apace into the world which only a few years ago was the realm of science fiction.

At a more philosophical level, the Japanese case provides supportive data for those who point out that we must pay attention to social death and to the concept of personhood. We strongly support those who urge that the central issues remain open for ongoing debate and discussion. Perhaps there can be general agreement upon some very broad parameters; beyond that we need flexibility that can be called into play for each particular case. North Americans are in general uncomfortable without closure, we do not tolerate uncertainty well. But surely an ability to be flexible has to be at the heart of any pluralistic medical system, and also at the heart of the ethics which inform that system.

NOTES

1. Brain death can be described as a state of irreversable coma which is now generally accepted as death in the West, although the body may be kept alive on a respirator, often for the purpose of organ removal for transplants and/or experimentation. This state is usually carefully distinguished from what is known at a "persistant vegetative state." This point is developed in the text.
2. These statistics were transmitted verbally by the Registry of the International Society for Heart Transplants.
3. An opportunistic sample of eighteen Japanese informants was used in this study. A reasonably exhaustive coverage was made of newspaper and magazine articles from 1982 onwards.
4. A Japanese collegue has stated in a personal communication that he has never heard of any evidence for this claim made by Feldman.
5. There have been some startling newspaper articles recently which indicate that the Japanese may be quite correct in their concern both about false diagnosis and financial exploitation in connection with organ donation. The *Montreal Gazette* reported a case in Ottawa in January 1989 in which a 79-year-old man who had been officially declared brain dead sat up in bed in response to the voice of his grandson and who is now, once again, leading a normal life. In London recently there have been questions in Parliament about kidneys that were bought from four Turkish donors and imported to England for use in a private hospital. There are also some very macabre reports by Amnesty International in connection with the sale of Bangladeshi children for slaughter and the extraction of parts of their bodies for export and eventual transplantation.

REFERENCES

Ad Hoc Committee of the Harvard Medical School to Examine the Definition of Death
    1968   A Definition of Irreversible Coma. Journal of the American Medical Association 205: 337-340
Amano, K.
    1987   *Nōshi o Kangaeru, Zōki Ishoku to no Kanren no Naka de* (Thoughts on Brain Death in Connection with Organ Transplants). Gekkan Naashingu 15, 13: 1949-1953.
Andrews, L.B.
    1986   My Body, My Property. Hastings Center Report 16: 28-38.

Asahi Shimbun
1987a *Nōshi Yoninha wa 23%* (23% Accept Brain Death).
1987b *Nōshi Kenkai de Funkyū Tsuzukeru Gakujutsu Kaigi* (Confusion over Brain Death Position Continues in Science Council of Japan).
1987c *Nōshi* (Brain Death). November 26.
1988a *Shinki Ishoku Shinpo Ryū wa Zōka* (Increase in Those in Favor of Accepting Progress in Heart Transplants). April 14.
1988b An Endless Variety of Images: The Penetration of Understanding in Connection with Brain Death is Essential. April 14.
Bartlett, Edward T. and Youngner, Stuart J.
1988 Human Death and the Destruction of the Neocortex. *In* R. Zaner (ed.) Death: Beyond Whole-Brain Criteria. Dordrecht: Kluwer Academic Publishers. Pp. 199–216.
Bernat, J.L. et al.
1981 On the Definition and Criterion of Death. Annals of Internal Medicine 94: 389-391.
Engelhardt, H.T.
1984 Allocating Scarce Medical Resources and the Availability of Organ Transplantation. Special Report, New England Journal of Medicine 311: 66-71.
Engelhardt, H.T.
1986 The Foundations of Bioethics. New York: Oxford University Press.
Feldman, E.
1985 Medical Ethics the Japanese Way. Hastings Center Report 15: 21-26.
Fox, R. and Swazey, J.
1973 Chronical of a Cadaver Transplant. Hastings Center Report 3: 1-3
Fuchs, V.R.
1974 Who Shall Die? Health, Economics and Social Choice. New York: Basic Books Inc.
Fukuda, M.
1975 A Survey Research of Doctors' Attitudes Toward Euthanasia in Boston and in Tokyo. Osaka University Law Review 22: 19-77.
Igakkai Shimbun
1987 *Naze Nihonjin wa Zōki Teikyō ni Shōkyokuteki na no ka* (Why are Japanese Unenthusiastic About Organ Donation?) October 26: 2-3.
Journal of the American Medical Association
1986 Spontaneous Movements in Brain-Dead Patients. Letters to the Editor 255: 2028
Katō, I.
1988 *Nōshi Mondai; Shakaiteki Gōi wa Shinkirō* (The Brain-Death Problem: Social Concensus is a Mirage). Bungei Shunju 4: 106-115.
Katō, I. et al.
1986 *Nōshi Zōki Ishoku to Jinken* (Brain Death, Organ Transplants, and Human Rights). Tokyo: Yuhikaku.
Law Reform Commission of Canada
1979 Criteria for the Determination of Death. Ottawa: Minister of Supply and Services Canada.
Lebra, T.
1976 Japanese Patterns of Behaviour. Honolulu: University of Hawaii Press.
Levy, R.I.
1973 Tahitians: Mind and Experience in the Society Islands. Chicago: University of Chicago Press.
Lifton, R.J. et al.
1979 Six Lives, Six Deaths: Portraits from Modern Japan. New Haven: Yale University Press
Lock, M.
1980 East Asian Medicine in Urban Japan: Varieties of Medical Experience. Berkeley: University of California Press.

1988   New Japanese Mythologies: Faltering Discipline and the Ailing Housewife. American
        Ethnologist 15: 43-61.
Lock, M. and Gordon, D. (eds.)
    1988   Biomedicine Examined. Dortrecht: Kluwer Academic Publishers.
Long, S. and Long, B.
    1982   Curable Cancers and Fatal Ulcers. Social Science and Medicine 16: 2101-2108.
Mainichi Daily News
    1984   Organs Removed from Woman Without Consent. December 24.
Mainichi Shinbun
    1988   *Shujutsu Manyuaru Kōhyō* (Making Public of the Operation Manual). February 28.
Nakajima, M.
    1985   *Mienai Shi: Nōshi to Zōki Ishoku* (Invisible Death: Brain Death and Organ Transplants).
            Tokyo: Bungei Shunju.
Namihira, E.
    1988   *Nōshi, Zōki ishoku, Gan Kokuchi* (Brain Death, Organ Transplants, Revealing a Diagnosis
            of Cancer). Fukumu Shoten.
Needham, J.
    1962   Science and Civilization in China. Vol. 2. Cambridge: Cambridge University Press.
Newman, B.M.
    1940   What is Death? Scientific American 162 (June): 336-337.
Pernick, Martin
    1988   Back from the Grave: Recurring Controversies over Defining and Diagnosing Death in
            History. *In* R. Zaner (ed.) Death: Beyond Whole-Brain Criteria. Dortrecht: Kluwer
            Academic Publishers. Pp. 17–74.
President's Commission for the Study of Ethical Problems in Medicine and Biomedical and
Behavioral Research.
    1981   Defining Death: Medical, Legal and Ethical Issues in the Determination of Death.
            Washington, D.C.: U.S. Government Printing Office.
Prottas, J.M.
    1983   Encouraging Altruism: Public Attitudes and the Marketing of Organ Donation. Milbank
            Memorial Fund Quarterly 61: 278-306.
Rado, L.A.
    1981   Death Redefined: Social and Cultural Influences on Legislation. Journal of Communica-
            tion 31: 41-47.
Rosaldo, M.
    1984   Toward an Anthropology of Self and Feeling. *In* Culture Theory: Essays on Mind, Self, and
            Emotion. Cambridge: Cambridge University Press. Pp. 137-157.
Shrader, D.
    1986   On Dying More than One Death. Hastings Center Report 16: 12-17.
Simmons, R. et al.
    1987   Gift of Life: The Effect of Organ Transplantation on Individual, Family, and Societal
            Dynamics. New Brunswick, N.J.: Transaction Books.
Somerville, M.
    1985   Access to Organs for Transplantation: Overcoming Rejection. Canadian Medical Associa-
            tion Journal 132: 113-117.
Starzl, T.E.
    1985   Will Live Organ Donations No Longer Be Justified? Hastings Center Report 15: 5.
Wartofsky, M.
    1988   Beyond a Whole Brain Definition of Death: Reconsidering the Metaphysics of Death. *In* R.
            Zaner (ed.) Death: Beyond Whole-Brain Criteria. Dordrecht: Kluwer Academic Publishers.
            Pp. 219–228.

Williams, J.R.
    1985   Human Organ Sales. Annals of the Royal College of Physicians and Surgeons of Canada
            18: 401-404.
Wright, P. and Treacher, A. (eds)
    1982   The Problem of Medical Knowledge: Examining the Social Construction of Medicine.
            Edinburgh: University of Edinburgh Press.
Yomiuri Shimbun Editorial Department
    1985   *Inochi Saisentan: Nōshi to Zōki Ishoku* (The Leading Edge of Life: Brain Death and Organ
            Transplants). Tokyo: Yomiuri.
Yonemoto, S.
    1986   *Bioeshikkusu-kō* (Thoughts on Bioethics). Karada no Kagaku 129: 6-9.
Young, A.
    1988   Unpacking the Demoralization Thesis. Medical Anthropology Quarterly 2: 3–16.
Youngner, S.J. et al.
    1985   Psychosocial and Ethical Implications of Organ Retrieval. Sounding Board. New England
            Journal of Medicine 313: 321-324.
Zaner, R.
    1988   Introduction. *In* R. Zaner (ed.) Death: Beyond Whole-Brain Criteria. Dordrecht: Kluwer
            Academic Publishers. Pp. 1–14.

PATRICIA A. KAUFERT

ETHICS, POLITICS AND CONTRACEPTION:
CANADA AND THE LICENSING OF DEPO-PROVERA

Depo-Provera, the trade name for depot medroxyprogesterone acetate, is a long-acting, injectable contraceptive. It is approved in Canada only for the treatment of endometriosis and as an adjunctive or palliative treatment for certain cancers. In 1984, Upjohn (the manufacturers of Depo-Provera) applied to Health and Welfare Canada (the ministerial department responsible) for a license to market Depo-Provera as a contraceptive for unrestricted use by Canadian women. Once publicly known, this application was opposed by a group of consumer, health, labour, union and women's organizations, who came together under the title, the Coalition on Depo-Provera. This essay provides a brief chronicle of the campaign against the licensing of Depo-Provera organized by that Coalition, but has as its main focus an exploration of the complex ethical and political issues which have been raised by the debate over this contraceptive.

These issues include those which are standard to bioethical debate, patient autonomy and informed consent. Viewing autonomy as an abstract principle, a bioethicist might argue that the characteristics and method of administering Depo-Provera do not in any way change the ethics governing physician behaviour. His/her responsibility to weigh patient autonomy against patient well-being remains the same. But seen from the perspective of a patient to be given Depo-Provera, this method of contraception produces a significant shift in the balance of control over her fertility away from the woman and toward the care-giver. Once outside the immediate context of her encounter with a physician, a woman can chose whether or not to take a contraceptive pill. But Depo-Provera is given by injection and lasts at least three months. Non-compliance ceases to be an option.

Informed consent is an issue partly because of the marked lack of information on Depo-Provera, particularly on the long-term risks associated with its use. Theoretically, consent is given or withheld after the patient has balanced benefits against risks and side-effects using the information supplied by the physician. But what if adequate information on risk is not available to either patient or physician? Can consent be termed "informed" under these conditions? Under what obligation is the physician to disclose that information on risk is inadequate? Should a physician allow his/her judgement of the immediate benefits to a patient to outweigh the fact there is little knowledge of the long-term risks?

The questions raised by Depo-Provera relative to informed consent and its impact on the balance of power and moralities of the physician-patient relationship are important, but in the tradition of bioethics, their focus is individualistic. As a social scientist, I question whether the discussion of ethics in medicine should be confined to this level of discourse. Hiller (1981) makes a distinction between medical ethics and health ethics, describing the former as concerned

G. Weisz (ed.), Social Science Perspectives on Medical Ethics, 121–141.
© 1990 Kluwer Academic Publishers. Printed in the Netherlands.

with "individual patient care issues," whereas health ethics is concerned "with broader health (and public policy) issues affecting the public at large" (Hiller 1981: 6). While this formulation is useful in extending ethical debate beyond the confines of the physician-patient encounter, the separation between the two forms of ethics, and between micro and macro levels of analysis, creates a false dichotomy. Some issues, such as the licensing of Depo-Provera, cut across these divisions; for example, the ethical sensibilities and dilemmas of the physician-patient relationship continually infiltrated the debate over the licensing of Depo-Provera. Yet, this debate also dealt with issues that went beyond whether or not Depo-Provera should or should not be included among the contraceptives commonly offered by Canadian physicians to Canadian women. For granting a license, or alternatively listening to the protests of the Coalition, had general implications for health policy and drug regulation in Canada. Lending Canadian government approval to Depo-Provera, or withholding this approval, held consequences for international family planning programmes and for women internationally.

Rather than finding a new label (such as health ethics) I would argue for a broader interpretation of medical ethics, one which includes issues of politics and policy. For I see the campaign against the licensing of Depo-Provera as a practical exercise in the political dimension of medical ethics. The aim of the Coalition on Depo-Provera was to deepen and broaden the debate over this contraceptive beyond the narrowly defined parameters set by government and industry by raising a series of questions: "Who should have a say in decisions for and against approval of Depo-Provera?"; "Should there be a place in drug regulation for women, the potential consumers of the drug, along with government, industry and medical experts?"; "How much consideration in this decision making should be given to the impact that approval/non-approval may have on women outside Canada?" These are questions about consent, but they occur at a macro rather than a micro level, and they are political.

When the Coalition asked whether the drug was safe for Canadian women, it was dealing with the informational aspects of consent. Their questions were broad ranging. The Coalition wanted to know about funding, about the conduct of research, about which risks are investigated and which not, about where research was done and why, about the implications (epidemiological and ethical) of using women of the Third World as test subjects, about the acceptability of generalizing from these women to women in Canada. As seen by physicians and health bureaucrats, these questions were irrelevant, ill-mannered or obtuse. Seen by a social scientist, these same questions touch critically on the ways in which epidemiological knowledge is (or is not) produced. Seen from within a feminist ethics, these are questions about the ethics (or lack of ethics) in research on women's health, the power of medicine, the powerlessness of women.

Reference in a volume on biomedical ethics to a feminist ethics requires explanation. As defined by Sherwin (1987), a feminist ethics sets ethical issues in the context of the social and political realities in which they arise, and resists the attempt to evaluate actions or practises in isolation (as traditional responses in bio-medical ethics often do) (Sherwin 1987: 279).

For feminists, whether philosophers or social scientists, questions of fertility and reproduction involve issues of power and patriarchy. Contraception is as much a political as a medical matter, it is also a matter of intense ethical concern.

The licensing of Depo-Provera has to be approached from within the frame-work of a feminist ethics, partly because this is an issue in women's health, partly because the campaign against it was conducted largely by women and women's groups, but primarily because a traditional bioethical approach cannot deal with issues raised by the Coalition but a feminist ethics can. For example, wherever Depo-Provera has been used, it is more likely to have been given to women whose fertility is devalued by their society. These tend to be women in poverty, women with disability, women from ethnic minorities, teenagers seen by health professionals as too young to be sexually active. Questions of autonomy and informed consent cannot be dealt with traditionally, in abstraction from these social and political realities. The actual injection occurs in the context of a physician-patient encounter, but often the reasoning for giving the injection has to do with the relative value placed on the rights of women to control their own bodies. For the strongest support for Depo-Provera has not been medical but comes from those who claim that the fertility of some individuals or groups should be curtailed in their interest, or in the interest of some wider group. Similarly, the motivating force for the Coalition was not medical but based on a belief in the absolute inviolability of a woman's body, whoever she may be, and wherever she may live.

These various issues will be dealt with in the second half of this essay; the first will establish the background for this drug and will summarize the campaign surrounding its approval in Canada. I must begin, however, by acknowledging that this account is partial in a double sense. It is selective because it is based on material in the public domain or made available to me by the Coalition on Depo-Provera. A more complete history would require access to information held by Health and Welfare, the Upjohn Company, and other interested parties to this debate, including the Committee on Reproductive Physiology, and the international family planning agencies. The essay is partial also in the sense of not being neutral. It is written in support of a particular perspective, that of the opponents of Depo-Provera. Finally, it is partial because I was myself involved actively in the campaign against Depo-Provera.[1]

BACKGROUND

Depo-Provera is not simply another new drug. The application for its approval arrived in Canada dragging behind it a long and controversial history, a history with which many of the actors party to the application process, (the Upjohn

Company, the bureaucrats of Health and Welfare, their medical advisors on the Committee for Reproductive Physiology, the Coalition) were familiar. Depo-Provera was developed by the Upjohn Company in the later 1950s as treatment for endometriosis and for threatened or habitual miscarriage. While unsuccessful for the latter purpose, women given the drug experienced a delayed return of menstruation and fertility. Almost inadvertently, the Company had produced a new type of contraceptive with an effectiveness equal to the pill, but given by injection, and with a single shot providing protection for at least three months.

Depo-Provera is currently licensed for sale as a contraceptive in over 90 countries (Richard and Lasagna 1987), but figures on how extensively it is used are uncertain.[2] Depo-Provera has several characteristics which make it a favorite of the international family planning agencies, particularly its long action, reversibility and ease of administration. Quicker than inserting an IUD, it is also without the concomitant risks of infection. The changes in menstruation which are its most obvious side-effect (spotting and amenorrhea) have been described by supporters as a positive advantage for women who are malnourished and overworked. (Planners also complain, however, that many women dislike the change in their menstrual pattern and discontinue their injections. Cf. Shain and Potts 1984.) Most importantly, unlike oral contraceptives, Depo-Provera does not depend on the continued compliance of women who might forget or abandon their pills under pressure from their families. From a programming perspective, Depo-Provera is efficient, relatively cheap and can be given by para-medical personnel; all these attributes make it an ideal contraceptive from the perspective of planners who want to provide coverage to large groups – such as women of the Cambodian refugee camps (Viravaidya et al. 1983; Greer 1984) – or living in remote areas serviced by an occasional mobile clinic – such as women living in the villages in northern Thailand (Ariwongse 1981).

Depo-Provera has been a controversial drug since studies in beagle dogs first suggested it might be a carcinogen. Supporters of Depo-Provera argue that little or no evidence exists of an increased risk of cancer to humans. Critics claim that there is no evidence that it is safe for women. The difference between these two positions seems slight, yet it lies at the ethical core of the debate over this contraceptive.

The history of Depo-Provera prior to the application for a license in Canada is partly a record of the remarkable persistence of Upjohn in trying to win or keep approval for its product. A request by Upjohn to market Depo-Provera for long-term use was rejected twice by the United Kingdom Licensing Authority; the company then asked for special hearings, got them and won. Successful in keeping Depo-Provera on the market after official inquiries in Sweden and West Germany, the company was less successful in Australia (Fraser 1987) and failed in successive attempts to get a license in the United States.

Not having its U.S. license has blocked Upjohn from the large domestic market and closed off direct access to U.S. international aid monies.

The State Department's Agency for International Development does not buy or fund the purchase of drugs for use overseas that are not approved by FDA, and some countries, such as India, do not sanction drugs that are unapproved in the country of origin (Sun 1984).

Furthermore, Upjohn's hold on its existing markets and its ability to increase sales of Depo-Provera internationally were continually threatened by accusations that the drug was unsafe (Senanayke and Rajkumar 1981). Inevitably, critics cited non-approval by the FDA as evidence. Upjohn finally asked the FDA for the appointment of a special Board of Inquiry and summoned up an array of medical experts to support Depo-Provera. Despite their testimony the Board decided that it was not satisfied that Depo-Provera was "safe for general marketing in the United States" (Weisz et al. 1984) and recommended that Upjohn should not get its license. At about the same time as this report was issued, the company is said to have applied for a license to market Depo-Provera as a contraceptive in Canada.

THE CAMPAIGN AND THE COALITION

At the time of Upjohn's application, Depo-Provera in Canada, although not approved, could be prescribed legally as a contraceptive.

> The Canadian Food and Drugs Licensing Act and Regulations at the federal level do not interfere with the right of physicians to use non-controlled drugs in a manner consistent with their sound medical judgement for their individual patients (Lucis 1984: 3).

Such decisions were "regarded as 'the practice of medicine'" (Lucis 1984) and came under the jurisdiction of the provincial Colleges of Physicians and Surgeons. Non-approval simply set "constraints on a pharmaceutical manufacturer's promotion and advertising of a drug" (Kinch 1982).

Upjohn's application for a license to market Depo-Provera as a contraceptive presumably coincided with the preparation of a report on "Oral Contraceptives" published by the Department of Health and Welfare in 1985. Prepared by the Special Advisory Committee on Reproductive Physiology, the report included a strong endorsement for Depo-Provera. Given the support of experts appointed to advise the minister on issues of reproductive health, Upjohn may have been confident that its application would go smoothly. The company seems also to have hoped that the drug regulatory process operated by a different set of mores in Canada than in the United States. Referring to the checkered career of Depo-Provera in the U.S.A., a spokesman for Upjohn was quoted as claiming that:

> We do things in a more private way in Canada. Down there, as soon as complaints about approval started coming in to senators and congressmen, they had to open up the

process. Here, it is really just a matter between us and Health and Welfare (*Globe and Mail*, November 22 1985).[3]

The process worked almost as Upjohn expected, but not quite privately enough. The news got out and an opposition formed. Many of the individuals and groups who came together in the Coalition on Depo-Provera had worked together previously in DES Action groups[4] and on the campaign to recover compensation for women injured by the Dalkon Shield. Some worked directly in international development programmes (such as Oxfam), or had links through the international women's movement with individuals and groups who were opposing the use of Depo-Provera in their own countries (Tudiver 1986). As a group, these women had political experience, a network through which to organize quickly and access information, and a deep distrust of official or medical arguments that a product was safe.

The Coalition declared that its objective was to "open up the process" and bring the question of approving Depo-Provera into the public and political arenas. The first stage was to get government to agree to public hearings, organized along the lines of the U.S. model decried as un-Canadian by the spokesman from Upjohn. The official response was moderately conciliatory, but the idea of hearings was initially rejected. At one point in the campaign, the Minister of Health was quoted in the press:

> If I do it [hold hearings] for Depo-Provera, I'd have to do it for all of them.... I am very loath to open the regulatory process (*Winnipeg Free Press*, August 9 1986).

Then, quite unexpectedly, Health and Welfare announced in June 1986 that a series of meetings would be held across Canada in September. The statement of objectives in the letter of invitation to participants proposed:

> To obtain views regarding contraceptive products and to provide information on various forms of contraception, including the status of Depo-Provera (Letter of Invitation).

Although officials from the department were careful to refer to "meetings" rather than "hearings," the Coalition believed that these meetings were a response to the pressures brought by women on the government.

Yet, these were not to be the type of public hearings for which the Coalition had been asking. Depo-Provera had become one method among many methods of contraception and one issue among many issues to be discussed. Critics were to be heard, but under tightly controlled conditions and with the minimum of publicity. In Winnipeg, for example, the media and anyone not invited as a participant were excluded. The number of presenters was limited and carefully mixed in terms of viewpoint. The amount of time allotted to each speaker was restricted and discussion tightly controlled.

The report on these meetings was another disappointment for the Coalition. Promised by Health and Welfare for the end of October, it did not appear until March 1987; it was a brief and cautious document with few references to the debate over Depo-Provera. It scrupulously avoided making any recommendations and listed nine "recurring issues," only one of which referred explicitly to Depo-Provera:

> Advocacy groups were adamantly opposed to the use of Depo-Provera as a contracep-
> tive while professional groups generally recommended it be provided as an option for
> carefully selected patients (Health and Welfare 1987).

From the perspective of the Coalition, the report was no more than a careful exercise in bureaucratic diplomacy, avoiding the major issues and skimming lightly over the criticisms of Depo-Provera made in the meetings. The Coalition itself was dismissively labelled as "advocacy groups" and reference to their opposition to Depo-Provera was bracketed in the same sentence with the "support of professional groups."

The Coalition complained to the Minister, again demanded public hearings and was again refused. Yet, for whatever reasons, Health and Welfare still did not move to give Upjohn its license and the Coalition was advised that the Minister was collecting further briefs and submissions.[5] Hence, when this essay was first written in May 1988, Upjohn had not got its approval, the Coalition had not got its public hearing, and the government seemed caught in a stalemate, deciding neither for nor against the licensing of Depo-Provera.

POLITICS, ETHICS AND DEPO-PROVERA

The brief chronicle of the campaign against Depo-Provera is a record of events occurring in the political rather than the medical arena, and seemingly well removed from the level of the physician-patient encounter. Yet, any decision by a government agency about any drug (not only Depo-Provera) is ultimately a medical decision, in the sense that it is implemented by physicians on the bodies of patients.

Earlier in this essay, I reproduced a quotation by a spokesman for Upjohn in which he referred to the drug regulation process in Canada as an affair solely of the manufacturer, department bureaucrats, and a committee of medical experts. This can be read both as a description of an administrative process, and as a statement about who has the right to participate in decisions about which drugs should be available and under what conditions. The Minister of Health clearly saw the decision over licensing of Depo-Provera as his responsibility. Replying in the House of Commons to a question from Gilbert Chartrand M.P., he said,

> I have said to health groups and to women's groups that, before any approval would be

given, if in fact a recommendation were made along that line, I would want to be satisfied totally on the safety side of the issue.

Members of the Committee on Reproductive Physiology may rather have agreed with colleagues in the United States, such as Schwallie (1984). Describing the FDA Board of Inquiry, Schwallie listed the various review committees that had endorsed Depo-Provera:

> The World Health Organization, the Committee on Safety of Medicines in the United Kingdom, the special ad hoc panel of the U.S. Government, the advisory committee of the Health Protection Branch, Canada, the Royal College of Obstetrics and Gynecology, and the American College of Obstetricians and Gynecologists.

He continued:

> In all instances, these learned bodies found that it was reasonable and prudent to continue the use of Depo-Provera. Responsible individuals, making responsible decisions, reached the conclusion that the drug is acceptable for use (Schwallie 1984: 569).

By implication, a government agency (such as the FDA and its Board of Inquiry) which did not follow the advice of its medical experts was irresponsible. The assertion that the decision over Depo-Provera is a purely medical matter can be seen as a tactic used in any battle for medical control. Wolf suggests that such a claim is an assertion that "The particular piece of turf belongs to physicians alone – that they are the appropriate ones to make the decision" (Wolf 1988: 200). In matters of contraception and through the lens of a feminist ethics, it is also a denial of the right of women to participate in decisions affecting their bodies.

The statement by the Upjohn spokesman made no allowance for the rights of the public (as exercised through consumer groups like the Coalition) to be involved in the decision over Depo-Provera. Seen from the company perspective, many of its difficulties with Depo-Provera are the product of the intervention of such groups; company bitterness over this issue is evident in a quotation from another Upjohn spokesman, taken from a speech made at a conference in Thailand:

> When the FDA did not add contraception to the list of approved indications for Depo-Provera, it renewed the social and political controversy over the use of Depo-Provera throughout the world. In addition to accusations leveled against the drug, rather vicious attacks are being directed against government agencies, international organizations and individuals throughout the world (Duncan 1981).

Supporters of Depo-Provera, such as Fraser in Australia, saw public protest as an intrusion into a purely medical matter. In a review of the history of Depo-Provera and the FDA Board of Inquiry for the Fertility Society of Australia, he continued:

> If looked at honestly from a scientific perspective it [the approval of Depo-Provera] should not be a matter of "public concern," but can only be so if the political and anecdotal emotive viewpoints are considered (Fraser 1987).

He added:

> I have reviewed this in some detail because it is a lengthy saga, some of it scientific, but much of it just inflexible argument between groups who have polarized themselves on emotional and political issues (Frazer 1987: 7).

Fraser is making two claims, the first is that the licensing of Depo-Provera is not a question in the public domain, but should be decided by the medically expert. The second claim, perhaps the more fundamental one, is that science is neutral, something separate from politics and emotion.

The place of values in science will be discussed later; for the moment I want to deal with the place of politics and emotion, but also of ethics, in the campaign against Depo-Provera. When women are told that the experts "know" about the safety of a drug (such as Depo-Provera) the emotional response is often one of anger. The folklore of the women's movement is full of stories of medical experts disproved, for it is a folklore which draws on the history of DES, the Dalkon Shield, Thalidomide. Indeed, the emergence of feminist health groups in Europe and the United States and Canada ran parallel with, and was partly in response to, a succession of disclosures about the risks of various hormones, including the contraceptive pill. The conviction that women have been victims is coupled with anger that harm has often been preventable, a product of carelessness and a lack of concern over the health of women; for example, describing the history of the IUD, Ruzek writes:

> Even fairly sophisticated critics of the health care system in the United States were shocked to learn that intra-uterine devices (IUDs), touted as extremely safe and effective, were never tested by the Food and Drug Administration or any federal regulatory body before being marketed (Ruzek 1986: 187).

A resolution among some women to prevent recurrence of such histories led into a concerted effort to influence public policy on matters affecting women's health, particularly in the area of hormones, reproduction and contraception (Ruzek 1986). The notion that the personal *is* the political was taken in a literal sense by women who argued that the defense of the body must be conducted in the political arena; but the personal also became the ethical. Women who are health activists believe they have a responsibility towards other women and an obligation to prevent another series of tragedies similar to those caused by DES or Thalido-mide.[6]

When news broke of the probable licensing of Depo-Provera in Canada, a political response was axiomatic for those Canadian women already familiar with the history of this drug in the U.S.A. and elsewhere. The methods used by the Coalition were intended to bring the maximum political pressure to bear on the government. These included press releases, radio and television interviews, and public meetings. A campaign of letters was organized; women wrote to the Minister of Health and to national and provincial politicians. Representations against Depo-Provera were made to Health and Welfare by concerned organizations working internationally (such as Oxfam) and by those working in Canada with women likely to be prescribed Depo-Provera (such as the Canadian Association for Community Living). Questions about Depo-Provera were asked in the House of Commons by Cyril Keeper M.P. (Canada Parliament, November 3 1985), Neil Young M.P. (December 2 1985) and Gilbert Chartrand M.P. (March 25 1986). Provincial politicians were asked to lobby their federal counterparts. The same tactics were used when the Coalition was pressuring the Minister for the release of the report from the meetings held across Canada. These are political practices, but the motivation has to do with a feminist interpretation of the two classic dilemmas of medical ethics: patient autonomy and informed consent.

When the Coalition demanded public hearings on Depo-Provera, it was challenging the whole drug-approval and drug regulatory process. Its claim was that women had rights not only to be informed, but also to participate, bringing forward their witnesses, presenting their own perspectives. At one stage in the campaign for public hearings, the Manitoba Minister responsible for the Status of Women issued a statement which included the following sentence:

> Canadian women have a right to full information on medications and the methods and criteria used to approve them.

The issue is essentially the classic one of informed consent, but as a right operant at the level of government decision making, not at the micro level of the individual patient.

DEPO-PROVERA, SAFETY AND THE PRODUCTION OF MEDICAL KNOWLEDGE

In the above quotation, the Minister claimed that women had a right to information about Depo-Provera, but also the right to know how this information was evaluated. This essay is not the place to undertake another review of the epidemiological literature. My concern is rather with the way in which the debate over Depo-Provera, and the disagreements over levels of risks and safety, are rooted in very different ways of looking at the production of medical knowledge.

Medical opinion on whether Depo-Provera is a carcinogen has been (is) divided. On the one hand, after reviewing the twenty or so epidemiological studies then available, the FDA Board of Inquiry concluded that as a general rule:

> Researchers did not include enough women, especially long term users, failed to follow users long enough to detect cancer, used controls that were inappropriate or inadequate, did not take into account a subject's particular risk for various cancers, and failed to record data systematically (Sun 1984).

The counter argument has been essentially the same as the one made by Schwallie (1984); namely, that other committees of medical experts had reviewed the same literature and were satisfied.

Supporters of Depo-Provera usually explain why these experts should disagree by attacking the Board of Inquiry for its overly strict interpretation of the rules of epidemiological evidence in the case of Depo-Provera (Fraser 1987; Schwallie 1984; Richard and Lasagna 1987). Schmidt (1981) made similar criticisms of the FDA as part of a general review of the politics of drug research and development in the United States. After reviewing the reasons why the FDA Board of Inquiry and its equivalent in the United Kingdom reached different conclusions on the licensing of Depo-Provera, Richard and Lasagna (1987) came to the conclusion that:

> The question is not whether policy shapes the data, but whether policy shapes the interpretation and weight of the data and thus the rendered judgements (Richard and Lasagna 1987: 886).

The difference between the British and the American authorities was largely that the former were looking for evidence that there was risk, whereas the latter were looking for evidence that Depo-Provera was safe. These differences in interpreting the rules of evidence are, as Tesh (1988) has pointed out, partly a matter of different traditions and customs. Experts are not neutral and the interpretation of scientific data is not value free. In her brilliant analysis of Agent Orange, Tesh wrote:

> Despite a desire to keep science separate from politics, scientists must turn to their values to help them define the pertinent facts. They have to decide whether they hope to prove dioxin safe or hazardous (Tesh 1988: 150).

Similarly, in the case of Depo-Provera, whether people wanted to prove that this contraceptive was safe or hazardous depended on their values, regardless of whether they were physicians, feminists, bureaucrats or company representatives.

Unlike clinical scientists, social scientists generally recognize the place and impact of values in research, including their own. They do not have a view of science as neutral, but see it rather as an activity embedded in its social and political context. Most would agree with Wylie that,

"Received view" ideals of objectivity and the conviction that the scientific method is self-cleansing are simply untenable; the products of science are marked by the identities of its practitioners, the social conditions under which they operate, and the cultural projects in which they engage (Wylie 1987: 63).

Tesh (1988) suggests that feminist writers, such as Sandra Harding (1986), take this same view of the social embeddedness of science, but that they argue also that:

When the dominant political ideology is racist, sexist, classist and imperialist, these assumptions will necessarily permeate scientific research (Tesh 1988: 176).

Once research on Depo-Provera is seen from this more radical perspective, the standard epidemiological questions about its quality, while still necessary, are not enough.

Questions asked by the Coalition involved looking at who has been doing the research. The answer was that epidemiological research on Depo-Provera has been done largely by the Upjohn Company itself, or by agencies with a major interest in international family planning programmes, AID, WHO, the Population Council and Family Health International. While eminently respectable, these agencies are family planning rather than disease focused. Feminists suspect them of playing down the risks of contraception for a woman in their effort to control her fertility (LaCheen 1986).

Money, and who has provided it, have largely determined how medical knowledge about Depo-Provera is produced, in what quantity, and for what purpose. The amount spent on any form of contraceptive research has been relatively stagnant since the 1970s. A review of how the available money is allocated reported that between 1980-83 only 7% to 8% of the total amount went to long-term safety evaluation; "expenditures on the safety of Depo-Provera and other injectables represented less than one percent" (Forrest et al. 1985). Money is not itself in short supply (witness the amounts spent on coronary heart disease or cancer), but the political priority given to investigating the risks associated with injectable contraceptives is low. The subjects are women, mainly poor women and women living largely in the Third World.

According to the Coalition, money has also determined what has been researched. Research on the relationship between cancer and Depo-Provera became a priority once Depo-Provera was labelled as a potential carcinogen. Little research money, or attention, has been expended on other potential disease outcomes (such as coronary heart disease,[7] stroke, gall bladder disease, diabetes, or abnormalities in the fetus). Yet, these are outcomes which have been associated with other hormones given to women. Information on reversibility, or such side-effects as break through or prolonged bleeding, is limited. The preoccupation with cancer reflects the emphasis on proving the absence of risk rather than the existence of safety. At the same time, the choice is also a matter of politics and economics. The suspected association between Depo-Provera and cancer was an impediment on the market. Other potential forms of risk, while acknowledged, did not threaten approval of the drug, and were ignored.

One of the more formal epidemiological criticisms of research on Depo-Provera is that results are not generalizable from Third World to Western women. Re-stated from a feminist perspective, research has not been grounded in the social and political reality of the women for whom Depo-Provera would be prescribed in Canada. Depo-Provera may increase the risk for diabetes, for example, but data are scarce. The lack of information takes on a new significance when seen in terms of Native Canadian women, who are both at high risk for diabetes and probable candidates for Depo-Provera. Depo-Provera is recommended for Canadian women over 35. Yet, the epidemiological data on cancer are based on women who are much younger and live in areas where the risk of some diseases, such as breast cancer is much lower than in Canada. The reality behind these results is that Depo-Provera may be prescribed in ignorance of the levels of risk associated with age and being Canadian. Information on the return of fertility is available only from a study done with relatively young women in Thailand. The grounds for promising reversibility to older women, or even to teenagers from the Canadian streets are insufficient. "Not being generalizable" means that we cannot properly know the risks.

Deciding on whether information is sufficient, applicable, of adequate quality, is not a value free exercise. Discussing the caution with which scientists interpret their results, Tesh (1988) argues that caution is itself not neutral.

> If an (inevitably) flawed study shows a substance to be toxic, it is cautious on behalf of the substance to reject the study, but cautious on behalf of the potentially exposed people to accept it. In contrast, if another (inevitably) imperfect study shows the substance to be benign, it is cautious on behalf of the potentially exposed people to reject it but cautious on behalf of the substance to accept it (Tesh 1988: 150).

Translated to the case of Depo-Provera and the arguments of the Coalition, informed consent becomes a question of arguing that caution should be exercised on the behalf of those women who might be given Depo-Provera, rather than on behalf of the Upjohn Company and the substance itself.

DEPO-PROVERA, AUTONOMY AND REPRODUCTIVE FREEDOM

Even were Depo-Provera safe, and even were the informational conditions for informed consent adequately satisfied, the issue of autonomy would remain. The right of women, whether as individuals or as a group, to control the reproductive power of their own bodies is central to the debate over Depo-Provera. Discussing broader issues of reproductive technology, Sherwin notes that traditional bioethicists [would] urge us to separate the question of evaluating the morality of various forms of reproductive technology in themselves from questions about particular uses of that technology. From the perspective of a feminist ethics, however, no such distinction can be meaningfully made. Reproductive technology is not an abstract activity, it is an activity done in particular contexts and it is those contexts which must be addressed (Sherwin 1987: 282).

Upjohn has always denied allegations that Depo-Provera is intended for poor women, or for "second class citizens." But in a commentary on the ethics of the debate over Depo-Provera in the United States, Levine (1979) suggested that it is difficult to avoid the conclusion that the drug's targets would be those or similarly vulnerable groups, such as institutionalized mentally ill women, and the mentally retarded.

When recommending Depo-Provera in its report on contraception, the Committee on Reproductive Technology listed the groups in Canada for whom they saw Depo-Provera as the most useful:

[Those] who smoke, especially if over the age of 35; [those] with a high risk of infection from IUDs; [those] at risk of cardiovascular problems, regardless of age and smoking habits; [those] who have had repeated failures when using other methods; [those] for whom amenorrhea would provide relief from dysmenorrhea or premenstrual syndrome; [those] for whom amenorrhea provides a hygienic or medical advantage; [those] who cannot tolerate the side effects of estrogens; [those] who are peri-menopausal (Health and Welfare Canada 1985: 110).

In essence, the Committee was recommending Depo-Provera for women for whom the pill or IUD is unsuitable, either because of their age, their medical status, or because they are seen as actually (or probably) non-compliant. The Coalition has argued that within this non-compliant category are found the promiscuous teenager, women belonging to ethnic and cultural minorities (particularly when there are problems of communication) and the severely mentally disadvantaged.

When arguing that particular groups of Canadian woman were more likely to be given Depo-Provera, the Coalition could point to existing information on the characteristics of women to whom physicians had prescribed Depo-Provera. Data from the U.S., Great Britain and New Zealand show that women in minority groups (whether they are Black and Navaho women in the Southern U.S., immigrant women in London (Savage 1982) or Maori women in New Zealand) are more likely to have been prescribed Depo-Provera. In a study of contraceptive users in Manawatu, New Zealand, for example, Trlin and Perry (1987) reported "a marked over-representation of women from the lowest socio-economic strata" and that 25% of the Maori women in their sample had been given Depo-Provera. Use of Depo-Provera with women living in institutions has been documented in state facilities for the mentally retarded in Tennessee, in government facilities for the mentally retarded in Ontario (Zarfas 1981) and in institutions for Navaho women run by the Indian Health Service. Thai authorities forced Depo-Provera on Cambodian refugee women (Greer 1984) and women in South Africa have complained that agreeing to a shot of Depo-Provera was a condition of being given work (Hartmann 1987). The Thai and South African cases are extreme examples; nevertheless, they are part of the general trend. Whatever Upjohn may claim, political and ideological factors, afford a better explanation of these patterns than do medical.

Noting the over-representation of Maori women among users of Depo-Provera, Trlin and Perry (1987: 32) suggest medical practitioners and family planning clinic staff have a poor mental and/or clinical image of the Maori and less educated women as reliable contraceptors. A similar attitude was reported among health professionals (physicians and nurses) in Manitoba who were interviewed as part of a small study undertaken by the Coalition on Depo-Provera (Hellsten 1987). Discussing the clients for whom they prescribed Depo-Provera, they talked about adolescents "with very unstable life style with no routine [who] are at high risk of sexually transmitted disease. They saw giving Depo-Provera to such adolescents as a way to 'buy them three months' to put their lives more in order" (Hellsten 1987). The commitment to these teenage clients came through as strong and genuine, but paternalistic. The underlying assumption was that if a teenage pregnancy did not make sense to the physician, the latter was entitled to impose this judgement on the teenager, if only for the period of time that the injection would last. The argument is that pregnancy and childbirth also carry risk. However, these risks are known, whereas those for Depo-Provera are not. More importantly, the risks feared tend to be social rather than medical.

LICENSES, CANADA AND THE ETHICS OF GLOBAL POLITICS

Pressure on the Department of Health and Welfare to license Depo-Provera came partly from clinicians who treated the opposition of the Coalition as an attack on the integrity of the medical profession. The physicians interviewed by the Coalition, for example, were sensitive to headlines in the press which had proclaimed that Depo-Provera was a banned and dangerous drug. While knowing that their own prescription of Depo-Provera was legal in the formal sense, they saw the granting of a license as a form of retrospective legitimization. Conversely, refusal of a license called continued use of Depo-Provera into question, not in a legal, but certainly in an ethical sense. One of the problems for the Coalition was to convince them that the issue had implications beyond Canada, and beyond the sensitivities of well-meaning physicians.

The Coalition believed that the reason for Upjohn's application was unlikely to lie in the appeal of the Canadian market, which is small for any contraceptive product (relative to the U.S.) and even smaller for Depo-Provera. An alternative explanation is that approval by Canada, a country with a high international reputation, may have been seen by Upjohn as a valuable counter to FDA disapproval, or even as a form of "back-door" entry into the U.S. market. If this interpretation is accepted, then the opposition to Depo-Provera in Canada becomes a matter of global not just national politics.

Mary O'Brien (1981) has called this "the age of contraception." Commenting on O'Brien's analysis, Held (1989) writes:

> The development of the technology by which women can control reproductive processes is in O'Brien's view what Hegel calls a world historical event. It can be expected to alter drastically the consciousness of women and relations between men and women (Held 1989: 382).

Contraception offers freedom to some women and the loss to other women of even the power to reproduce. Discussing the development of contraceptive technology, O'Brien suggests that its development had little to do with the liberation of women but:

> more likely, [with] concern about the capacity of high technology to keep the population of the world employed, worry about political instability in the Third World and among the unemployed in the bastions of capitalist production, and the loss of control of capital itself in the energy crisis (O'Brien 1981: 205).

The risks of this age of contraception are experienced by all woman, but particularly by women whose fertility is devalued by their own and Western society, the women of the Third World.

The counter argument is that the risks associated with childbirth are so high for countries of the Third World, that the risks of contraception are a matter of lesser concern. Physicians who have worked in refugee camps, or in countries where poverty and malnutrition are rife, fully believe that Depo-Provera saves lives that would be lost under the repeated impact of pregnancy and childbirth. They defend contraception in general, and Depo-Provera in particular, as in the interest of women's health.

Support for Depo-Provera comes also from those who see fertility of Third World women not in terms of their health, but rather as a threat to economic and political order. They have a strong sense of imminent disaster as a consequence of overpopulation. These views are expressed by Hammerstein writing in the *American Journal of Obstetrics and Gynecology*. After complaining about FDA refusal to approve Depo-Provera, Hammerstein continued:

> There is no time to lose – it is less than 5 minutes till midnight. If the young generation, which constitutes nearly 50% of the population in some developing countries, behaves as their parents did after starting sexual life and reproductive activity, the problem of overpopulation will become disastrous. All developmental progress would be negated, with malnutrition, pauperization, and epidemic diseases being inevitable conse-quences.... We all should try to prevent this development, even if this means swimming against a torrential stream (Hammerstein 1987: 1023).

Hammerstein's fears are widely shared, from the World Bank through the various international agencies concerned with family planning.

Feminists have very different views of population control and family planning, and these are summarized in the following statement by Levine:

> The best results in population control do not depend on family planning programs alone, and certainly not on the availability of one contraceptive, but on a whole range of measures that improve income distribution, raise the educational level of women, provide better maternal and infant health care, better sanitation, more jobs, and more opportunities for meaningful choices in life. Given that range of social and economic measures, women will seek on their own to control fertility, and thus motivated, will choose methods that are best suited to their medical and other needs (Levine 1979).

For most feminists, this statement represents the only ethical position available to women. It is also the philosophy which guided the Coalition on Depo-Provera.

CONCLUSION

This essay is not a sociological study of the Coalition. Nor is it intended as a neutral discussion of one ethical position relative to another. It is about the importance of political commitment. It is about the need for an activist approach in ethics. Finally, it is about the ethical dimension of relationships between women as much as or even more than it is about medical ethics in the traditional sense. Sherwin writes:

> A feminist ethics is a moral theory that focusses on relations among persons as well as on individuals. It has as a model an inter-connected social fabric, rather than the familiar one of isolated, independent atoms.... It is a theory that focusses on concrete situations and persons and not on free floating abstract actions.... It is a theory that is explicitly conscious of the social, political, and economic relations that exist among persons; in particular, as a feminist theory, it attends to the implications of actions or policies on the status of women (Sherwin 1987: 279).

For the most part, women at risk of being given Depo-Provera are not the Canadian women of the white middle class. The women at risk live in the Third World, or are Canadian women who are poor, Native, immigrant, the mentally, physically or morally disadvantaged in the eyes of the community. It could be argued that Depo-Provera should not concern those not at risk, but only if one denies Sherwin's model of an inter-connected social fabric as the basis of relationships between women. The anger of women, indeed my own anger and sense of commitment to this issue, is grounded in a strong sense of obligation to all women. It is an anger rooted in an ethical obligation to oppose harm, whether actual or potential to the bodies of women; it is an anger based on taking moral offense.

EPILOGUE

A few developments have occurred since this paper was first written. Suspecting that Upjohn was still actively lobbying for approval, the Coalition has maintained its own campaign, writing regular letters to the Minister, keeping a watching brief on the latest literature and research on Depo-Provera, holding the occasional press conference, maintaining international ties with other groups opposed to this contraceptive.

The most important developments have been in the research field with the publication of two papers. The first was based on a study in Costa Rica and reported a significant association between Depo-Provera and breast cancer (Lee et al. 1987), but not cervical cancer. (Slightly earlier results from a WHO study had described a higher, although not statistically significant, relationship between Depo-Provera and cervical cancer, but not breast cancer.) The second paper came out of Thailand, a centre for research on Depo-Provera, and reported a significant association between Depo-Provera use by the mother and polysyndactyly and chromosomal anomalies in the infant (Pardthaisong et al. 1988). Neither of these studies are conclusive, but both add to the climate of suspicion that surrounds Depo-Provera. The Coalition wrote to the Minister calling his attention to both papers and also discussed them at a press conference held by the Coalition in December, 1988. Replying to the Coalition, the Minister announced that Upjohn had been asked to provide the results of additional safety studies before further consideration could be given to its application. This announcement may or may not mark the end of the campaign to license Depo-Provera in Canada, as well as of the Coalition opposing it.

NOTES

1. I reviewed the epidemiological literature relevant to the proposed prescription of Depo-Provera to women aged thirty-five and above.
2. Speaking before a committee of the U.S. Senate in 1987, a spokesman for the American College of Obstetricians and Gynecologists claimed that approximately 1.5 million women were current users of Depo-Provera (Morley 1987). In the same year, the Chairman of the WHO Steering Committee on Long Acting Agents for Fertility Regulation, told the Fertility Society of Australia that: "There are probably five million current users at a conservative estimate, and 250,000 long term users – that is, women using it for four, five years or more" (Fraser 1987).
3. The same article reported that the Director of the Human Prescription Drug Branch had said that approval was expected in December, 1985.
4. Diethylstibestrol (DES) was given to woman with high risk pregnancies to prevent miscarriages. It was later found to cause a rare form of vaginal cancer in their daughters. DES action groups were formed in many countries, including Canada with the objective of informing the public about the dangers of DES and helping those who had been exposed to its risks.
5. The Canadian Nurses Association was among the groups approached and was to decide against supporting the approval of Depo-Provera later that year.
6. The National Women's Health Network presented briefs on Depo-Provera to the FDA Board of Inquiry; groups equivalent to the Canadian Coalition have been (are) active in Australia (Bunkle 1984) and the U.K. (Rakusen 1981).

7. After reviewing the data on the relationship between Depo-Provera and coronary heart disease and finding them sparse, the Board of Inquiry appointed by the U.S. Food and Drug Administration commented: "Clarification of these issues is especially important, in particular since one of the proposed indications for the use of DMPA is for subjects with other risk factors for cardiovascular disease, such as hypertension, which might be exacerbated by estrogens" (Weisz et al. 1984: 152).

8. Sweden has licensed Depo-Provera for domestic use, but the Swedish International Development Authority decided not to deliver or finance the purchase of Depo-Provera in any of its programs or projects in the Third World, because continuous medical follow-up is unlikely to be available.

REFERENCES

Ariwongse, Boonchom
  1981   Mobile Family Planning Clinic, McCormick Hospital. *In* Second Asian Regional Workshop on Injectable Contraceptives. Oklahoma: World Neighbors. Pp. 54-56.

Bunkle, Phillida
  1984   Calling the Shots? International Politics of Depo-Provera. *In* Rita Arditti et al. (eds.) Test-Tube Women: What Future for Motherhood? London: Pandora Press. Pp. 165-187.

Canada Parliament
  1985, 1986 House of Commons: Debates; Official Reports. Ottawa.

Duncan, Gordon W.
  1981   Medical/Legal Aspects of Depo-Provera. *In* Second Asian Regional Workshop on Injectable Contraceptives. Oklahoma: World Neighbors. Pp. 45-50.

Forrest, Darroch et al.
  1985   Worldwide Trends in Funding Contraceptive Research and Evaluation. Family Planning Perspectives 17, 5: 196-208.

Fraser, Ian
  1987   Depo-Provera: A Matter of Public Concern? Special Report. Addressed to the Fertility Society of Australia International Conference on November 12.

Greer, Germaine
  1984   Sex and Destiny: The Politics of Human Fertility. Great Britain: Secker and Warburg.

Hammerstein, Jürgen
  1987   Contraception: An Overview. American Journal of Obstetrics and Gynecology 157: 1020–1023.

Harding, Sandra
  1986   The Science Question in Feminism. Ithaca: Cornell University Press.

Hartmann, Betsy
  1987   Reproductive Rights and Wrongs. New York: Harper & Row.

Health and Welfare Canada
  1985   Report on Oral Contraceptives by the Special Advisory Committee on Reproductive Physiology to the Health Protection Branch, Canada.
  1986   Regional Meetings on Fertility Control. Health Protection Branch: Winnipeg, Manitoba, September 8.

Held, Virginia
  1989   Birth and Death. Ethics 99: 362-388.

Hellsten, Cathy
  1987   Study of Depo-Provera Use in Manitoba. Preliminary Report. Paper presented at the 11th Annual Conference of the Canadian Research Institute for the Advancement of Women, Winnipeg, November.

Hiller, Marc D.
  1981   Medical Ethics and Public Policy. *In* Marc. D. Hiller (ed.) Medical Ethics and the Law: Implications for Public Policy. Cambridge, Mass.: Ballinger. Pp. 3-45.

Kinch, Robert
    1982   Should Depot Medroxyprogesterone Acetate be Considered for Additional Uses? Canadian
            Medical Association Journal 127: 947.
LaCheen, Cary
    1986   Population Control and the Pharmaceutical Industry. *In* K. McDonnell (ed.) Adverse
            Effects: Women and the Pharmaceutical Industry. Toronto: Women's Education Press. Pp.
            89-136.
Lee, Nancy C. et al.
    1987   A Case-Control Study of Breast Cancer and Hormonal Contraception in Costa Rica. JNCI
            79, 6: 1247-1254.
Levine, Carol
    1979   Depo-Provera and Contraceptive Risk: A Case Study of Values in Conflict. Hastings Center
            Report 9, 4: 8-11.
Lucis, O.J.
    1984   Medroxyprogesterone Acetate (Depo-Provera). Canadian Perspective 1983/84. Presented
            to The Special Advisory Committee on Reproductive Physiology, February 10.
Manitoba Information Services
    1986   Minister Urges Release of Depo Provera Report, December 2.
Morley, George W.
    1987   Statement of The American College of Obstetricians and Gynecologists on Depo-Provera
            Use as a Contraceptive. Presented before the Subcommittee on General Oversight and
            Investigations Committee on Interior and Insular Affairs, U.S. House of Representatives,
            August 6.
O'Brien, Mary
    1981   The Politics of Reproduction. Boston: Routledge and Kegan Paul.
Pappert, Ann
    1985   Disputed Drug May be Approved as Contraceptive. The Globe and Mail, November 22.
Pardthaisong, T. et al.
    1988   Steroid Contraceptive Use and Pregnancy Outcome. Teratology 38: 51–58.
Rakusen, Jill
    1981   Depo-Provera: The Extent of the Problem – A Case Study in the Politics of Birth Control.
            In H. Roberts (ed.) Women, Health and Reproduction. London: Routledge and Kegan Paul.
            Pp. 75–108.
Richard, Barbara and Louis Lasagna
    1987   Drug Regulation in the United States and the United Kingdom: The Depo-Provera Story.
            Annals of Internal Medicine 106: 886-891.
Ruzek, Sheryl
    1986   Feminist Visions of Health: An International Perspective. *In* Juliet Mitchell and Ann
            Oakley (eds.) What is Feminism? New York: Pantheon Books. Pp. 184-207
Savage W.
    1982   Taking Liberties with Women: Abortion, Sterilization and Contraception. International
            Journal of Health Services 12: 293-306.
Schmidt, Alexander
    1981   The Politics of Drug Research and Development. *In* Henry Wechsler, R.W. Lamont-Havers
            and G.F. Cahill (eds.) The Social Context of Medical Research. Cambridge, Mass.:
            Ballinger. Pp. 233-262.
Schwallie, Paul C.
    1984   Depot Medroxyprogesterone Acetate Update. *In* G. Zatuchni et al. (eds.) International
            Workshop On Long-Acting Contraceptive Delivery Systems. New York: Harper and Row.
            Pp. 566–80.
Senanayake, Pramilla and Renuka Rajkumar
    1981   Answering Public Criticism on Depo-Provera. *In* Second Asian Regional Workshop on
            Injectable Contraceptives. Oklahoma: World Neighbors. Pp. 74-83.

Shain, Rochelle N. and Malcolm Potts.
1984 Need for and Acceptability of Long-Acting Steroidal Contraception. *In* G. Zatuchni et al. (eds.) International Workshop On Long-Acting Contraceptive Delivery Systems. New York: Harper & Row. Pp. 1–19.

Sherwin, Susan
1987 Feminist Ethics and In Vitro Fertilization. *In* Marsha Hanen and Kai Nielsen (eds.) Science, Morality & Feminist Theory (Canadian Journal of Philosophy, Supplementary Volume 13). Calgary: University of Calgary Press. Pp. 265-284.

Sun, Marjorie
1984 Panel Says Depo-Provera Not Proved Safe. Science 226: 950-951.

Tesh, Sylvia Nobel
1988 Hidden Arguments: Political Ideology and Disease Prevention Policy. New Brunswick and London: Rutgers University Press.

Trlin, Andrew and Paul Perry
1987 Depo-Provera Use In the Manawatu User Characteristics and Implications. New Zealand Population Review 7: 28-34.

Tudiver, Sari
1986 The Strength of Links: International Women's Health Networks in the Eighties. *In* Kathleen McDonnell (ed.) Adverse Effects: Women and the Pharmaceutical Industry. Toronto: Women's Educational Press. Pp. 187-215.

Viravaidya, Mechair, Arunee Fongsri and Thomas D'Agnes
1981 The Role of Contraception in Community-Based Programs. *In* Second Asian Regional Workshop on Injectable Contraceptives. Oklahoma: World Neighbors. Pp. 22-25.

Weisz, J., G.T. Ross and P. Stolley
1984 Report of the Public Board of Enquiry on Depo-Provera. Rockville, Maryland: Food and Drug Administration.

Winnipeg Free Press
1986 Epp Refuses to Budge: Drug Hearings to Remain Closed, August 9.

Wolf, Susan M.
1988 Conflict Between Doctor and Patient. Law, Medicine & Health Care 16: 3-4.

Wylie, Alison
1987 The Philosophy of Ambivalence: Sandra Harding on the Science Question in Feminism. *In* Marsha Hanen and Kai Nielsen (eds.) Science Morality & Feminist Theory (Canadian Journal of Philosophy, Supplementary Volume 13). Calgary: University of Calgary Press. Pp. 59-76.

Zarfas, D.E., I. Fyfe and F. Gorodzinsky
1981 The Utilization of Depo-Provera in the Ontario Government Facilities for the Mentally Retarded: A Pilot Project. Paper presented to the Ontario Ministry of Health, October.

PART III

THE INSTITUTIONS AND IDEOLOGY OF MEDICAL ETHICS

GEORGE WEISZ

THE ORIGINS OF MEDICAL ETHICS IN FRANCE: THE INTERNATIONAL CONGRESS OF *MORALE MÉDICALE* OF 1955

In 1955, the Ordre National des Médecins, the professional body responsible for regulating the French medical profession, organized what was in effect the first international congress devoted to medical ethics.[1] Although the term *éthique médicale* was already in circulation, organizers preferred to use the more traditional and less philosophical term *morale médicale* to describe the subject of the meeting. For congress organizers, this term was itself something of an abbreviation for an even wider series of issues, "professional medical regulation, medical morality and comparative medical law" (I: 7).

The very breadth of the subjects invoked indicates that the emerging concern with medical ethics among the French medical elite of the 1950s was part of a much wider configuration of preoccupations. Holding together these diverse preoccupations was the determination to defend traditional and highly individualistic notions of social relations in medicine. A close examination of the proceedings of this Congress thus demonstrates how the institutional domain of medical ethics in France was constructed by the medical elite in the 1950s in response to several converging pressures and dilemmas.

The first was the creation during the Vichy period of the Ordre des Médecins invested with responsibility for deontology and medical ethics. This created the first and most significant institutional framework for medical ethics. The second factor was growing concern with clinical testing. The routinization of human experimentation was troubling to many, particularly in the aftermath of the still-recent Nazi medical atrocities. Third, the long process begun in the 1920s to create a national health insurance system in France reached a critical phase in the 1950s. The result was intense political and social conflict. In this context, the medical elite began to feel considerable anxiety about the future of the physician's therapeutic autonomy, seen as the foundation for traditional medical ethics.

ORIGINS

For a large part of the 19th and early 20th centuries, medical opinion in France was agitated by the issue of whether or not to create a disciplinary body to regulate the medical profession. By the 1920s, a wide-ranging consensus in support of some such body was in place. But it was not until 1941 that the Ordre National des Médecins was finally established by the Vichy government. In spite of the questionable political conditions of its establishment, the Ordre survived the liberation. One of its first and most important post-war acts was the promulgation in 1947 of a deontological code binding on the French medical

G. Weisz (ed.), Social Science Perspectives on Medical Ethics, 145–161.
© 1990 *Kluwer Academic Publishers. Printed in the Netherlands.*

profession. This replaced an earlier non-binding code published by the Ordre in 1941. The new focus on deontology led to an intensified concern with the issues of medical ethics that had been discussed sporadically during the interwar period, most notably in the Catholic medical journal, the *Cahiers Laennec*.

The principal author of the code of 1947 was Louis Portes, Professor of Obstetrics at the Paris Faculty of Medicine and President of the Ordre des Médecins since 1943. Having many questions and holding even more opinions about the moral and political dilemmas of modern medicine, Portes took a characteristically French action. He went for help to the national academy that was supposed to decide issues of morals and politics, the Academy of Moral and Political Sciences.[2] This Academy had a rather checkered past and lacked the prestige and authority of comparable institutions in other domains, notably the Academy of Sciences and the Academy of Medicine. For one thing, the issues it discussed were less amenable to final judgement than were matters of science and technology. For another, the moral and political sciences were not easily subject to centralization in the French manner. The Academy's membership did not correspond to any single recognizable professional elite in contrast to the memberships of the Academies of Sciences and of Medicine.

Nevertheless, with the encouragement of the writer-physician, Georges Duhamel (himself an influential member of this Academy), Portes in 1948 addressed the Academy of Moral and Political Sciences on the moral issues facing modern medicine (Portes 1954: 105-112). A few months later, the Academy created a permanent commission to examine problems of professional ethics and deontology. During the next few years, medical practitioners gave talks on a variety of subjects including, euthanasia, medical confidentiality, patient consent, human experimentation, reproductive planning and the independence of the medical practitioner. The Academy occasionally acted on these presentations. In 1949, it condemned under any circumstance the deliberate provocation of death by a physician. Soon after, it recommended against artificial insemination where the husband was not the donor. Following the passage, in 1952, of a resolution affirming the utility of comparative medical law, the Ordre des Médecins collaborated with the Law Faculty of Paris to establish a center for the study of medical law.

Meanwhile another institutional expression of the medical elite, the Academy of Medicine, was also concerning itself with matters of medical ethics largely as a response to the behavior of German doctors during the Second World War. In 1949, the Academy named a special commission to consider a suggestion that French medical schools introduce a course in international medical ethics and that France support the establishment of a special medical section of the Permanent Court of International Justice. The Commission decided against taking unilateral action but did authorize its representatives to bring these issues up in the World Health Organization and the International Red Cross.[3] On several occasions during the 1950s, the Academy discussed the problem of ensuring medical neutrality in wartime.[4]

Nazi atrocities also provoked a more sustained concern with human experimentation. In 1952, the Academy devoted several meetings to discussing this issue in closed session. The Academy made public its conclusions at the end of 1952, deciding that 1) experimentation in the interests of the health of a patient was permissible when usual methods of diagnosis and therapeutics were not effective; 2) experiments on healthy humans were permissible only when there was no other method of resolving an issue, the subjects were volunteers exercising real freedom to choose, and tests were directed by a medical expert.[5] The Ordre also consulted the Academy on a number of other controversial issues including therapeutic sterilization.[6]

Medical ethics was being discussed in other contexts as well during these years. The *Cahiers Laennec*, published by an association of Catholic physicians, continued to be active in the postwar era. It devoted two special issues to human experimentation (1952) and single issues to such topics as artificial insemination (1946), abortion (1946), the code of deontology (1947), euthanasia (1949) and medical secrecy (1950). It was, however, the Ordre which did most to publicize and give institutional reality to medical ethics. In 1954, while it was in the process of preparing a revised deontological code, the Ordre, decided to publish a collection of essays and speeches on medical ethics by Louis Portes who had died five years before (Portes 1955). In that same year, it took even bolder action by deciding to organize a Congress of *Morale Médicale*. The date chosen was meant to commemorate the 10th anniversary of the ordinance of 1945 regulating the practice of French medicine. The gathering would take up many of the issues that had already been raised in the Academy of Moral and Political Sciences, in the Academy of Medicine and in the debates surrounding the revised deontological code. It would give the Ordre des Médecins an international audience for its discussions and was in a position to shape the intellectual framework within which its concerns were expressed.

THE PRELIMINARY REPORTS

Before the Congress convened, the Ordre published a volume of preliminary reports that were to constitute the framework for the meeting. The subjects were varied, in keeping with the organizers' understanding of the term *morale médicale*. The reports were divided into three parts. All those in the first two were written by individuals who were either members of or close to the national council of the Ordre des Médecins. This probably accounts for the uniformity of attitude underlying the different subjects discussed. Papers in the third section, by contrast, were prepared by jurists.

The first section was devoted to "professional medical jurisdiction" and sought to describe recent French experiences in this domain. Two substantial essays described respectively the functions of the Ordre des Médecins and the French system of specialist certification. Two minor papers discussed medical oaths and the Ordre's efforts to promote mutual aid among practitioners.

The heart of the volume was a second section devoted to *morale médicale* and comprised of two separate parts of almost equal length. The first contained three papers dealing respectively with medical confidentiality, patient consent and the independence of the physician. What was listed in the Table of Contents as a fourth report was actually a second series of three essays. These dealt respectively with the ethical conditions for new procedures in diagnostic medicine, for those in surgery (by far the longest), and for clinical testing of new medications. Thus the first series of papers in this second section dealt with fairly traditional issues relating to the doctor-patient relationship; the second was devoted to the impact of new technologies and clinical experimentation.

The final section was made up of two reports by jurists describing the results of surveys conducted by the Centre for Medical Law affiliated with the Law Faculty of Paris. One was devoted to the legal conditions for the practice of medicine in different countries and the other described legislation relating to medical confidentiality in different nations. Both were highly descriptive and were based on the written responses of individuals in twenty-two countries to a questionnaire. The authors of these two reports readily admitted that the data were far from comprehensive.

Somewhat surprisingly, the Congress excluded from its definition of medical ethics reproductive issues like abortion, sterilization and artificial insemination which were subjects of continuing public debate. It also excluded euthanasia which was being discussed even though it had few defenders in the aftermath of Naziism. It is likely that for medical as opposed to public opinion, these were not particularly troubling issues.[7] The subjects defined as ethical reflected the preoccupations of the organizers. The Ordre des Médecins was a powerful regulatory body whose attributions included the conditions permitting the practice of medicine, deontology, professional ethics *per se*, and relations with public authorities. These were the concerns which found their way into the preliminary papers.

There is however a more profound reason for the joining of issues of professional jurisdiction and regulation with those that we would now consider medical ethics. For spokesmen of the medical elite, the traditional foundations of medical practice *and* of medical ethics were being threatened by evolving institutional structures as well as by the development of new technologies. Perhaps the best statement of this attitude came from a relatively objective observer, Julliot de la Morandière, the former dean of the Paris Faculty of Law and president of the Academy of Moral and Political Sciences. In his opening address to the Congress, he described the malaise which had led doctors to turn to his Academy for aid (II: 34-36).

Since the French Revolution, he suggested, medicine had been a liberal profession. The state imposed strict conditions of access but medical practice remained free. The relationship between physicians and patients was based on traditional rules of morality and conscience dating back to Hippocrates: the confidence of the patient in the physician, a confidence supported by professional secrecy; the right of patients to independently choose a physician; and the full autonomy of the physician to choose his means of investigation and therapeutics.[8]

No one, said Julliot, was challenging these fundamental principles directly but their application had become increasingly problematic for two reasons. Firstly, scientific and technical developments had provided physicians with far more power to heal than ever before. Because the physician now wielded so much potentially dangerous power over the patient, his traditional independence of action was being called into question. The second reason had to do with the increasing costs of the new medical techniques. In order to bring the benefits of modern medicine to the entire population, costs had been assumed by the state. What remained of the physician's autonomy and professional confidentiality when civil servants interposed themselves between doctor and patient? Such preoccupations were, in fact, repeatedly expressed in the papers commissioned by the Ordre des Médecins for the Congress.

One could plausibly argue that the new concern with medical ethics in the 1950s represented a professional strategy for preserving medical autonomy. There is undoubtedly truth in this line of argument. The work of revising the deontological code that was then in process had provoked a major controversy over the wording of a phrase designed to protect the fundamental principles of traditional liberal medicine against state encroachment.[9] (The controversy probably explains the fact that the deontological code which was promulgated several months later was not mentioned at the Congress.) But this line of argument tells only part of the story and seriously oversimplifies the motives of the historical actors. For one thing, there existed areas of professional conflict that were far more heated. The generalization of the national system of health insurance after the war led to major battles between doctors and the state administration over such issues as fixed fee schedules and full-time employment of hospital physicians (Hatzfield 1963; Jamous 1966; Jamous and Peloille 1970). These conflicts were already building up to a fever pitch when the Congress convened and undoubtedly contributed to the general unhappiness with state medicine that was expressed. But they were not addressed directly by participants because they were not viewed as particularly *ethical* concerns.

Therapeutic autonomy, in contrast, had an entirely different status. Properly understood, in fact, autonomy *was* ethical behavior; medical ethics was neither possible nor conceivable without it. One need not accept the premises of this reasoning to recognize how deeply felt it was and how much it shaped French discussions of medical ethics.

The views of the medical elite are perhaps best approached through one of the preliminary essays that was devoted to the independence of the physician (I: 89-116). One of the two authors was J.R. Debray, then secretary-general of the Ordre des Médecins and the primary organizer of the Congress; the second, was Raymond Villey, a future President of the Ordre. It may or may not be significant that the essay was placed almost exactly in the centre of the published volume of preliminary papers (it appeared seventh among the twelve essays). But it was certainly at the heart of the values being defended at the Congress.

The defense of professional autonomy rested on a vision of medicine which insisted on the absolute centrality of the doctor-patient relationship. Each partner brought something specific to the exchange. The patient *chose* the physician and thus manifested faith in the healer's competence and moral qualities. The physician, in contrast, brought not only his expertise but his independence from all outside influences. Independence was thus defined as absolute concern for the well-being of the patient. There were many forms of dependence which relegated the patient's well-being to secondary status. Such dependence could be political, as the Nazi atrocities had shown. Medical independence in this context meant resisting political authorities and various emblematic heros were periodically cited as exemplars of this virtue. (Not the least of these were Louis Portes and, then Minister of Health, Bernard Lefay, both credited with forbidding doctors in 1944 to deliver wounded resistance members to the authorities.) In an age of national health insurance, the physician could also be constrained in his therapeutic actions by the regulations of bureaucrats. Finally, the physician's independence could be limited by his own self-interest, economic or otherwise, which might lead to acts that were not in the best interests of the patient.

Consequently, independence understood as devotion to the well-being of the patient provided the basis for medical ethics as well as for that patient confidence that was indispensible to medical care. But it was a very fragile basis indeed. It needed to be protected by codes of conduct and by the disciplinary jurisdiction of bodies like the Ordre des Médecins (which is why discussions of professional jurisdiction were entirely appropriate to a meeting devoted to ethics). Ultimately, however, it depended on the character of the medical practitioner which could not be legislated or even taught.

> That is because independence is not merely a juridical notion but also a behavior. It supposes firmness, energy, authority, self-confidence, a certain arrogance, perhaps. The exercise of a liberal profession very certainly develops these tendencies (I: 102).

Since function developed character, there existed real danger that physicians trained in the developing institutions of collective medicine would be lacking in the characteristics which made professional independence possible.

Debray and Villey, like many other speakers at the Congress, admitted the need for the collective and social medicine that was emerging everywhere. But they sought to safeguard the fundamental values of liberal medicine. It was normal, they stated, for third-party payers and sanitary authorities to try to restrict the independence of physicians which caused them no end of inconvenience. But doctors owed it to their patients to resist. What the authors feared most was that doctors would be motivated by their desire for economic security to capitulate without struggle. It was up to doctors and to the Ordre des Médecins to control the direction in which medicine was to evolve, to balance the needs of collective medicine with those of physician independence.

This radically individualistic notion of medical ethics owes its specific form and vocabulary to a current of Catholic humanism that was popular during the 1930s. (All the essential ideas in the essay by Debray and Villey seem to have been secularized and watered-down versions of those expressed in Okinczyk 1936.) But the sentiments expressed traditional and widespread ethical views. For contrary to a popular myth, medical ethics before bioethics was not just about professional etiquette. Much of it was about the protection of the patient. But this protection was to be achieved by ensuring that the physician behaved always altruistically and in the patient's best interests. The implicit model was the priest whose effectiveness depended on superior knowledge *and* total selflessness. The role of ethics was to bring about such behavior through exhortation or appeals to reason or by suggesting new institutional arrangements that might ensure proper behavior by physicians.

Traditional medical ethics was thus not about choosing between conflicting ethical imperatives. The one exception was an issue which often involved just such choice and which consistently received enormous attention in French medical and legal writings: medical confidentiality. (Neither "confidentiality" nor "secrecy" convey the full portentous implications of the French term *le secret médical*.) This was featured prominently at the Congress for this reason and, more importantly, because it epitomized the traditional view of the therapeutic relationship: the total confidence of the patient in the absolute devotion of the physician to his well-being despite all other claims. Two preliminary reports and four papers read at the Congress were devoted to confidentiality. It was moreover regularly invoked in other papers. Secrecy, said Debray and Villey, was as indispensable as the physician's independence.

> These two notions, independence and secrecy, are on the same level. They are the two
> fundamental pillars of medical ethics (I: 95).

The authors likened medical confidentiality to the secrecy of the confessional: "In such a domain divulging [information] would appear as a veritable betrayal of personhood." It was the basis of that medicine of the "individual person" which had, according to the authors, provided the foundation for medicine since antiquity.

The challenge to confidentiality did not stem from the fact that specific legislation could override it, as in the case of compulsory reporting of certain diseases. That was for the authors (as for the Ordre des Médecins) a legitimate instance of the primacy of collective well-being over that of the individual within well-defined limits. It was rather the growing desire for information of third-party payers of medical costs, whether public or private, that was viewed as the source of major ethical dilemmas. These were faced with a considerable degree of intellectual ingenuity by speakers at the Congress who nonetheless remained committed to the principle of medical secrecy.[10]

The premise that ethical behavior was altruistic and even charitable behavior, explains another prominent characteristic of discussions at the Congress: the lack of concern with patient autonomy. A role understood primarily in terms of providing relief to suffering humanity did not foster much concern for patients' rights and informed consent. Speakers were far more concerned to protect patients from unnecessary anxiety and anguish than to keep them well informed. Recognition of the importance of patient morale and faith in the healer in promoting cure also fostered tendencies that we would today characterize as paternalistic.

Patient consent was the subject of a preliminary paper and was brought up in a variety of other reports and communications. No one disputed the need for patient consent for medical procedures but there was some question about what this meant in practice. As a response to recent litigation, in fact, the legal counsel to the Academy of Medicine had recently asked that body to discuss the possibility of requiring the signature of forms before major medical procedures, if only to absolve doctors of any further legal responsibility.

If a patient vigorously refused a procedure there was no question, assuming his mental competence, that consent was lacking and that the procedure could not take place. The physician had, according to the Deontological Code, an obligation to attempt to convince the patient of the need for the procedure he wished to perform, but no more than that. In normal cases, suggested the authors of the major report on the subject (I: 75-88), J. Vidal and J.P. Carlotti, the physician elicited consent by explaining the medical situation and possible options. He did not however tell the patient everything. As a result of illness, the patient was presumed to have reverted to a child-like state in which full comprehension of the medical facts was not possible. Any information that would frighten or depress was to be studiously avoided. Doctors in France, as in other countries[11] generally had few doubts about the importance of the patient's psychological state for the medical outcome and thus routinely avoided communicating bad news to patients. Based as it was on quite limited information, consent was often far from complete. Nor did it have to be. Once again, it was the unique nature of the physician-patient relationship which ensured that consent was present.

Consent, argued Vidal and Carlotti, occurred at the moment that the patient chose his physician and placed his confidence in him. That is why the concept of free choice of practitioners was so critical. The ideal therapeutic situation in fact would be passive acquiescence to a doctor's orders. Since a patient's consent could never be completely free or enlightened, it was ultimately based on confidence in the physician. For that reason, the authors rejected the notion of signed consent forms. Even though these would afford physicians considerable legal protection, they would seriously limit the physician's liberty of therapeutic action especially in cases evolving over a long period. Physicians had to be prepared to respond to any eventuality "without discussion or delay." Signed forms would also cast a shadow of suspicion and mistrust that could not be therapeutically beneficial. A generalized consent form covering any eventuality would only duplicate unnecessarily the reality of the doctor-patient relationship. The consequence of this position was the principle that it is not the responsibility of the physician to prove consent but, in cases of conflict, that of the patient to prove lack of consent. Vidal and Carlotti cited with approval several recent court decisions which seemed to support this position.

The authors concluded their report with the reflection that there is nothing particularly attractive or agreeable for the healer in assuming the moral responsibilities implicit in this view of consent. Rather, it is the legal protection of forms, the freedom from choice which constitutes the more desirable alternative. But acceptance of limits to therapeutic freedom would amount to a betrayal of the mission to effect cures.

> To the extent he [the physician] escapes from the constraints which surround him, he finds, together with a large liberty of action, an increase in responsibilities and personal risks which are the ransom of a greater efficacy (I: 87).

The physician was, in other words, sacrificing his own tranquility and peace of mind to the therapeutic interests of the patient. His independence was a burden and a risk required by the nature of his task.

This approach to patient consent was developed further in the preliminary reports on the implications of new medical technology, particularly the one concerning surgery whose main author was Lortat-Jacob, another future President of the Ordre des Médecins (I: 125-166). Starting from the premises developed by Vidal and Carlotti, Lortat-Jacob asked how consent would work in the case of a cancer patient who required a mutilating procedure. There would be no problem in the case of someone who knew of his condition since the physician could simply explain the reasons for surgery. However, the more common situation in France was that patients were, for humanitarian reasons, not told that they had cancer. Much of the time the surgeon could obtain consent, the author believed, by evoking a vague future threat of the disease or by suggesting that a presently benign tumour was capable of turning malignant. In cases where radical mutilation was called for, there was probably no alternative to full disclosure. A refusal of consent by a fully-informed patient posed no particular moral problem for the physician. However, if the refusal came from a patient ignorant of his true condition the dilemma was "atrocious." The physician had to choose between not acting which might have serious medical repercussions or revelation which could demoralize the patient and affect his prognosis.

In the end, claimed Lortat-Jacob, the surgeon had to be a subtle psychologist capable of "affective diagnosis" as well as diagnosis of the surgical variety.

> Overall, even more than the positive and formal consent of the patient, the surgeon

must attempt to obtain a climate of confidence sufficient to reassure the patient and to prevent his asking too many questions (I: 134).

The case of lobotomy (or "psycho-surgery" in the language of the day) provoked different sorts of problems. (The authors assumed that its efficacy though not proven was sufficiently evident to justify it under specific conditions.) Sometimes the patient had the discernment necessary to give his consent; if he did not, consultation with the family might be possible. In the end, however, "it is the conscience of the physician, his scientific and moral value which should determine everything" (I: 155).

Aside from the problem of consent, the reports on new technologies had to do with the dangers associated with their experimental status and, less frequently, with their effects on the quality of the patient's life. The reports repeated the conclusions of the Academy of Medicine to the effect that fairly radical procedures including those at the experimental stage were permissible if they were being done for the benefit of the patient. Decisions of this sort most often boiled down to a weighing of risks and consequences of different actions. In every case, it was stressed that only the informed conscience of the physician could make such decisions. No bureaucratic organism could limit that freedom. Even the opinion of peers could not be imposed. There was to be something like absolute freedom of conscience for the physician in the name of the patient's best interests. Lortat-Jacob was aware that there was something "exorbitant" about this kind of claim but he insisted that "experience proves that the search for truth and progress, thus the alleviation of illness can only be obtained at this price" (I: 165).

Perhaps the most interesting essay in this section had to do with clinical testing of medications. It approved such testing on patients with illnesses that might be cured by the procedure; more controversially, it agreed with the principle of using healthy subjects. Such tests had to be preceded by all necessary laboratory experimentation, required the consent of the subject (imperfect though that consent was), were not to provoke illnesses in subjects who were well (which was proscribed in any event by French law) and were not to expose subjects to any grave risks. Above all, tests needed to be in the hands of medical experts.

This expert offers the patient the guarantee of his responsibility and his moral quality. Free from all constraints, he has for guides in the execution of his task only his conscience and his experience. Between him and his patient there are none of those chains of intermediaries who dilute responsibility to the point where the patient's rights are forgotten (I: 187).

Of vital importance, ethically speaking, was that healthy volunteers for experiments had to act on the basis of total liberty of action. This was understood so narrowly as to exclude or at least raise serious questions about many sorts of volunteers. The authors, for instance, considered it reprehensible for an experimenter to exploit the enthusiasm of members of his staff who might be so hopeful of a major discovery that they would willingly face serious danger. More pertinently, the definition of liberty of action excluded those who volunteered in order to obtain some advantage: dispensation of military service, money, reduction of prison sentences, hope of honor. At the very least, these were not very desirable collaborators. This definition of a test subject's liberty of action as essentially disinterested action is not unlike the definition of the physician's independence as disinterested care. There was a certain consistency to the ethical reasoning of the medical elite even if the definitions utilized were unlikely to occur frequently in real-life situations.

THE CONGRESS

On 30 September 1955, in the presence of various political and religious dignitaries, the Congress was convened in the great amphitheatre of the old Faculty of Medicine. According to the published proceedings, 450 individuals including 120 foreigners representing 26 nations participated officially at the meeting which lasted four days.

An effort was made to extend the scope beyond medicine with jurists granted particular prominence. There were two committees of patronage, one medical and another representing law, the social sciences and philosophy. (In reality, civil servants and jurists predominated.) Three jurists were listed on the organizing committee. Among those presenting introductory remarks were two jurists speaking respectively in the names of the Faculty of Law and the Academy of Moral and Political Sciences. This non-medical presence, however, served largely as window-dressing. Few of the individuals who gave presentations were without medical credentials. With only one exception to be discussed below, little in the way of a non-medical perspective was added to the two preliminary essays on comparative medical legislation.

In addition to the twelve already-published preliminary papers which were read and which provided the thematic structure of the meeting, nearly forty mostly brief papers were presented, about half of them by foreigners. An effort seems to have been made to give voice to some of the participants in earlier ethical debates who did not represent the official views of the Ordre des Médecins. Henri Baruk reiterated earlier views opposing any experimentation on healthy humans and condemning psycho-surgery (II: 123-130). The prominent academician Edouard Rist took issue with what he saw as an excessive French preoccupation with medical secrecy (II: 148-164). H.-P. Klotz presented an interesting critique of traditional notions of physician autonomy (II: 245-249). And in a talk which tried to take seriously the medical elite's insistence on the importance of patient morale in the process of healing, J. Hamberger insisted on the necessity of reforming hospitals in order to render them less oppressive (II: 110-114). Overall, few of the short papers presented at the Congress seriously challenged the views expressed in the preliminary essays commissioned by the Ordre des Médecins.

There were, however, two exceptions. Directly challenging the overwhelmingly pro-natalist sentiments of the French medical elite, the well-known birth control advocate, Mme. Lagroua Weill-Hallé defended contraception in one of the few papers to provoke strong disagreement (II: 181-187). Even more jarring were some informal comments by Professor R. Savatier of the Law Faculty of Poitiers, author of the preliminary essay dealing with comparative legislation regulating medical practice. At one of the sessions, Savatier expressed dismay at the paternalistic opinions being expressed which justified dissimulation and the telling of lies to patients. Consent based on such misinformation was not valid he declared. He also suggested that patients need not be treated as incompetent minors. The ideas of Portes which were the basis of the code of deontology and which resulted in the subservience of patients were now out of date. Savatier predicted that unless the medical profession came to understand the psychology of patients, it would eventually face a patient revolt.[12]

Savatier's remarks expressed the enormous distance between the experience and cultural values of the medical elite and legal thinking based on the notion of a contract between doctor and patient. His views were treated only briefly and in the most elliptical manner in the proceedings of the conference (II: 203). But they provoked an immediate defence of dissimulation from three medical professors who insisted on the cruelty in many situations of complete truthfulness which resulted only in anguish and terror. The few medical journals which reported on the Congress[13] were also struck by Savatier's comments. The editorialist of La Presse Médicale viewed the Congress as the confrontation of two tendencies: that of medicine could be accused of opportunism but attempted at least to respond to human need as it emerged in daily contact with patients; the juridical sought to transcend the human condition, as well as all spontaneity, in favor of the realm of law, itself variable and subject to the vagaries of politics and economics.[14] For another medical editorialist, the disagreement reflected the theoretical approach of jurists without personal responsibility to and direct experience of suffering humans.[15] Debate continued when Savatier developed his views more fully for a medical audience in an article appropriately entitled "At The Junction of Two Humanisms" (Savatier 1956). The exchange can be seen as one of the first conflicts of disciplinary values stemming from efforts to extend discussions of medical ethics beyond the medical profession.

The preliminary volume began with issues of professional organization which led into the ethics of the doctor-patient relationship in which the theme of physician autonomy was paramount. Only then did it discuss the problems arising from technology and experimentation. The Congress itself, however, started with the reports dealing with technology and experimentation before dealing with those concerning the doctor-patient relationship. The crucial essay on the independence of the physician by Debray and Villey was relegated to the end of a third section on professional organization.

One does not want to make too much of matters of scheduling which, any conference organizer will attest, can hinge on fairly trivial matters. But the consequence of this reordering was that the key issue of physician autonomy from regulation remained implicit in many of the early papers (which focused on the ethics of experimentation) until it was fully developed near the end of the meeting. The Congress thus began with the issues that were most agitating public opinion and climaxed with a defense of physician autonomy. The brief anonymous introduction to the second volume of conference proceedings also emphasized the dangerous depersonalization of medicine resulting from technology without introducing the theme of medical autonomy. Little wonder that no less an authority than the daily *Le Monde* viewed the meeting as a response to technology and human experimentation. The newspaper declared in the opening paragraph of a long summary article that the Congress demonstrated "the unanimous concern of doctors, jurists, moralists, from all countries, of all religions, to safeguard the 'human' in the medical act in the face of invading technology, in the face of audacious therapeutic tests."[16]

At the final session of the Congress, a motion was passed setting up a commission to examine all the motions and suggestions raised and to develop plans for an international society of medical ethics and law. Three weeks later, about thirty Congress participants met and passed a resolution stating their intention to create an International Society of Medical Ethics and Law. So far as I have been able to determine, such an association was not established. No agreement could be reached on a resolution concerning the moral conditions regulating the application of new medications and medical technology. Aside from the difficulties of achieving consensus, the Committee quoted with approval the words of the President of the Ordre des Médecins in inaugurating the Congress to the effect that it might be premature and dangerous for the Congress to come to firm conclusions on many issues and that the goal should rather be the exchange of ideas at the international level. This refusal to legislate proper behavior was fully consistent with the often-repeated principle that the conscience of the individual physician must be the final arbiter in difficult situations. The Committee did manage to pass one other resolution: this affirmed the need to preserve the independence of the physician as one of the essential characteristics of a civilization (II: 345).

CONCLUSION

The First Congress of *Morale Médicale* was unquestionably a public relations success. But its most concrete result turned out to be a resolution affirming the independence of the physician. One cannot escape the impression that this result was itself a consequence of the fact that the Congress was less an exercise in examining the new problems facing medicine than a reaffirmation of the validity of traditional solutions.

The medical ethics articulated at the Congress of 1955 revolved around issues of therapeutic power. This was not the first or last time that the language of ethics has been used to discuss and partially obfuscate issues of power relations. But the views expressed were not just about the struggle for power. They expressed a genuine moral vision which once had wide currency in western societies, though it seems anachronistic and even self-serving today. In this vision, the determining finality in the doctor-patient encounter was the improved health of the individual patient. It was most emphatically not his moral autonomy or some vision of collective welfare. Training and daily experience in a highly competitive milieu powerfully inclined doctors toward this moral vision. The inclination was further strengthened by the cold war atmosphere of the 1950s and by the evolving struggle over the conditions of national health insurance. Organizers of the Congress were convinced that they were grappling with fundamental matters of medical morality. They dealt with these issues, as we all do, on the basis of principles and assumptions that reflected their social class, professional experience and cultural values.

As in other nations, the rhetoric of medical individualism mellowed with time. The Second International Congress of *Morale Médicale* which took place in Paris in 1966, was full of references to the collective responsibilities of doctors and to the need for their participation in the elaboration of national policies (Ordre des Médecins 1966). But medical individualism has not lost its influence in France. At least part of the reason lies with the continued domination of medical ethics by the Ordre des Médecins, with relatively little input coming from other domains. The culture of the French medical elite, combining competitiveness, a sense of intellectual superiority and authoritarianism, has always been particularly consistent with an extreme form of individualism.[17]

It is at least plausible to argue that the power of this individualistic vision of medical ethics has been partly responsible for what has been perceived as the relative slowness of the French in developing coherent legislation on such matters as human experimentation (Fagot-Largault 1985: 105-148; Isambert 1987: 409). Similarly, the strong attachment to the idea that experimentation is only legitimate if it improves the health of the individual patient may lie behind the difficulties which the French have had in generalizing clinical testing. Traditional medical attitudes toward patient autonomy and informed consent also seem to have changed little since the 1950s.[18] Finally this individualist vision has provided the medical elite with little incentive to include the public in its ongoing discussions of ethical issues (Isambert 1983). It remains to be seen whether the political intervention which in 1983 created a more widely representative National Committee on Medical Ethics, has in fact changed matters in any fundamental way.[19]

ACKNOWLEDGEMENTS

Research for this essay was supported by a grant from the Social Science and Humanities Research Council of Canada. I am grateful to Donna Evleth and Elsbeth Heaman for helping me to collect the material on which this paper is based.

NOTES

1. The proceedings of this Congress were published in two volumes, Ordre National des Médecins 1955. Since this source will be cited so frequently, references in the text will include only volume and page numbers.
2. In tracing Portes' relationship with the Academy I rely on the account by Julliot de la Morandière in Ordre National des Médecins 1955, II: 36-37 and the preface by J.P. Carlotti to Portes 1954: 3-5.
3. *Bull. Acad. Nat. Méd.*, 1949, 133: 231-240, 278-279.
4. *Bull. Acad. Nat. Méd.*, 1953, 137: 274-277, 279-280. *Ibid.*, 1954, 138: 375-379. *Ibid.*, 1955, 139: 402-404.
5. *Bull. Acad. Nat. Méd.*, 1952, 136: 562-563.
6. *Bull. Acad. Nat. Méd.*, 1953, 13: 389-395. *Cahiers Laennec*, 1956, 16: 35-36.
7. Abortion and euthanasia were illegal and to most physicians unethical. Therapeutic abortion was troubling to many but it was hoped that technological progress would eventually make it unnecessary. It is possible, however, that a dispute about therapeutic abortion between the Catholics of the *Cahiers Laennec* and the Ordre des Médecins that had just surfaced in the former journal (1955, 15; 1956, 16) made this a delicate subject for an international meeting of this sort.
8. Traditional though they may have been, the actual formulation of these principles dates from the 1920s when the medical profession was forced to confront the prospects of national health insurance. Julliot seems to have been paraphrasing the wording of the Charter elaborated in 1927 by the Confederation of Medical Trade Unions. Interestingly, however, he neglected to mention the fourth principle of the Charter pertaining to direct payment of physicians by patients.
9. The issue seems fairly arcane in retrospect. Before the promulgation of the code of 1947 which contained an article stating that the four principles of liberal medicine were to be respected by all physicians, the Council of State had insisted that the article be qualified by a phrase stating that this requirement applied only in cases where its observation was not incompatible with legislative or regulatory prescriptions or where it would not compromise the normal functioning of the institutions and services of social medicine. In preparing its revised code, the Ordre sought to limit the applicability of this qualification by adding its own qualifications. In its wording, derogation was only possible if it could be "demonstrated" that the applications of liberal principles would harm social medicine and that such derogation did not damage the quality of medical care. The Confederation of Medical Trade Unions found this to be too moderate and proposed a wording that only allowed derogation of liberal principles where specific legislation was in conflict with these principles. The debate is chronicled in the issues of the *Bulletin* of the Ordre throughout 1955.
10. A consensus seems to have emerged at the Congress in favour of the notion of "shared secret." This meant that a physician could share information with another physician representing the insurance administration who was himself bound by confidentiality. This solution found its way into the deontological code promulgated shortly after the Congress. A useful summary of French thinking on the issue is Villey 1986.
11. The literature on the beliefs of American physicians during the 1960s is summarized in Taylor 1988: 441-442.
12. *La Presse Médicale*, 1955, 63: 1430. *Le Concours Médical*, 1955, 15 October: 3884. This was not the first time Savatier had made this point to a medical audience. Remarks published in *Le Concours Médical*, 1951, 5 April: 1307, had already created a stir (Planche 1956: 229-230).
13. Out of twelve major professional journals which I examined, only four covered the Congress. This helps place medical ethics in perspective among the preoccupations of the French medical profession in the 1950s.
14. *La Presse Médicale*, 1955, 63: 1430.
15. Dr. P.N. 1955: 1287.
16. *Le Monde* 1955, 6 October. The thematic progression from technology to autonomy was not perfect since the speech at the start of the Congress by the writer-physician Georges Duhamel was an extended attack on the inroads of the state bureaucracy on medicine (II: 43-46). Since Duhamel

(1955) wrote a piece on the Congress for *Le Figaro*, it is not surprising that this conservative newspaper took greater notice in its reports of the Congress' defense of professional autonomy than did the more leftist *Le Monde*.

17. Some aspects of this individualism early in the 19th century are discussed in Weisz 1987. The career structures of the medical elite in the 19th and 20th centuries are discussed in Weisz 1988.

18. Both these points are nicely illustrated by Isambert 1987: 409 who points out in his review of the literature relating to human experimentation that American works identify as problematic the issue of full consent by the subject whereas French works view as problematic testing of healthy individuals.

19. For some preliminary considerations on the workings of the Committee see Isambert 1986 and Moulin 1988.

REFERENCES

Bulletin de l'Académie Nationale de Médecine
    1949-1956 vols. 132-140.
Bulletin de l'Ordre des Médecins
    1952 no. 4.
Cahiers Laennec
    1946-1956 vols. 6-16.
Concours Médical
    1955 15 October.
Dr. P.N.
    1955 La morale médicale. La Vie Médicale 36: 1287.
Duhamel, Georges
    1955 Retour et triomphe d'Hippocrate. Le Figaro, 5 October: 1.
Fagot-Largault, Anne
    1985 L'Homme Bioéthique. Paris: Maloine.
Hatzfield, Henri
    1963 Le Grand Tournant de la Médecine Libérale. Paris: Les Éditions Ouvrières.
Isambert, François-André
    1983 De la bio-éthique aux comités d'éthique. Études 358: 671-683.
    1986 Révolution biologique ou réveil éthique? In Éthique et Biologie (Cahiers S.T.S.). Paris: Centre National de la Recherche Scientifique. Pp. 9-41.
    1987 La bio-éthique à travers ses écrits. Revue de Métaphysique et de Morale 92: 401-421.
Jamous, Haroun
    1966 Sociologie de la Décision; La Réforme des Études Médicales et des Structures Hospitalières. Paris: Centre National de la Recherche Scientifique.
Jamous, Haroun and B. Peloille
    1970 Change in the French University and Hospital System. In J.A. Jackson (ed.) Professions and Professionalization. Cambridge: Cambridge University Press. Pp. 111-152.
Le Monde
    1955 Un évenement d'une grande portée: Le Premier Congrès de Morale Médicale. 6 October.
Moulin, Anne Marie
    1988 Medical Ethics in France: The Latest Great Political Debate. Theoretical Medicine 9: 271-285.
Okinczyk, Joseph
    1936 Humanisme et Médecine. Paris: Labergerie.
Ordre National des Médecins
    1955 Premier Congrès International de Morale Médicale, I, Rapports; II, Comunications, Compte Rendu. Paris: Masson.
    1966

Deuxième Congrès International de Moral Médicale, I, Rapports; II, Communications, Compte Rendu. Paris: Ordre National des Médecins.

Planche, Henri
1956 Verité ou mensonge. Concours Médical, 21 January: 229-230.

Portes, Louis
1954 A la Recherche d'une Éthique Médicale. Paris: Masson and Presses universitaires de France.

Presse Médicale
1955 Chroniques: Premier Congrès International de Morale Médicale. 63: 1429-1431.

Savatier, René
1956 Au confluent de deux humanismes; entente et mésentente entre médecins et juristes. Concours Médical, 7 January: 61-74.

Taylor, Kathryn M
1988 Physicians and the Disclosure of Undesireable Information. In M. Lock and D. Gordon (eds.) Biomedicine Examined. Dordecht: Kluwer Academic Publishers. Pp. 441-464.

Villey, Raymond
1986 Histoire du Secret Médical. Paris: Editions Seghers.

Weisz, George
1987 The Self-made Mandarin; The Éloges of the French Academy of Medicine, 1824-1847. History of Science 70: 13-40.
1988 Les transformations de l'élite médicale en France. Actes de la Recherche en Sciences Sociales 74: 33-46.

MARGARET STACEY

THE BRITISH GENERAL MEDICAL COUNCIL
AND MEDICAL ETHICS

In this paper I propose to discuss the relationship between the General Medical Council (GMC) and medical ethics in the United Kingdom in the following way. First, I will describe briefly for those who are unfamiliar with the GMC what that body is, what its powers and functions in relation to medical ethics are and something about how it carries them out. Here I will include a note about the circumstances of my research on the GMC. Second, I will describe some of the changes that have taken place in the ethics guidance which the Council has given to registered medical practitioners in the UK over the past decade. I will then attempt to show the sources of these changes, specifically whether they have derived from inside or outside the profession. In the final section, I will suggest that the Council in providing guidance has moved closer to the concerns of the general public, but that it is still concerned to a large extent with intra-professional regulation. I will also suggest that this guidance seems nonetheless removed from many of the day to day ethical concerns of both practitioners and patients and that this is connected with variations in ways of knowing and understanding ethical problems as well as with differences of structural position.

GENERAL MEDICAL COUNCIL: STRUCTURE AND FUNCTIONS

The General Medical Council is the statutory body in the United Kingdom which regulates the medical profession. The Council is entirely financed by fees primarily those paid by registered practitioners; it receives no income from the state or from outside the profession. It is governed by the Medical Act of 1983 which consolidated a number of pieces of earlier legislation including the major reforms made in the Act of 1978.

The Council is responsible for the maintenance of the registers of those medical practitioners who are appropriately qualified. This involves both determining what credentials persons require before they may be admitted to the register and ensuring that those whose names remain on the register are fit to be there. The Council undertakes the first task by determining which medical schools provide education and examination to a standard such that their graduates may be said to be appropriately qualified. In connection with the second task it publishes the "blue pamphlet" *Professional Conduct and Discipline: Fitness to Practise* which offers guidance on issues to do with conduct and ethics. It is specifically guidance and not a code, on the grounds that the latter could never cover all cases. The blue pamphlet "was first published by the Council in 1963 ... when it superseded earlier and briefer Notices dealing with professional misconduct" (GMC 1978: 18).

The Council has, since a decision of 1979, had a Committee on Standards of

163

G. Weisz (ed.), Social Science Perspectives on Medical Ethics, 163–184.
© 1990 *Kluwer Academic Publishers. Printed in the Netherlands.*

Professional Conduct and on Medical Ethics (GMC 1981: 18). This is the permanent successor to a Special Committee which was established in 1975. The Standards Committee is responsible for doing all the preliminary work on the blue pamphlet, but it is full Council which has to implement any changes in the guidelines.

The Council has power to discipline registered practitioners who are shown to have fallen short of those standards or who have been convicted in the Courts. The Council handles discipline through a preliminary screening process done by or on behalf of the President. Such cases as the President or his deputy may decide, go to a Preliminary Proceedings Committee, which in its turn, passes certain of them to a Professional Conduct Committee. These Committees have a series of sanctions, the ultimate of which is erasure from the Register. Erasure, however, is not necessarily so final an act is it may seem, for practitioners may apply for reinstatement after ten months – some are restored and some not, then or later. In addition, since the 1978 Act, cases of doctors who are thought to be unfit to practise by reason of physical or mental ill health come under a different procedure, whereby they may submit to voluntary examination and surveillance until they are deemed to be sufficiently recovered to be fit to practise. If they refuse to comply voluntarily, they may be referred by the President or his nominee to a Health Committee or to a more informal health procedure.

The Council in the period 1976-1984 was composed as shown in Table 1. This is essentially its present composition, save that the Council increased the number of lay members by two in 1987. Any great increase in the number of lay members would also require an increase in the number of elected medical members or decrease in the number of nominated medical members since the Act requires that elected members should always have a majority on the Council.

The increase in the size of the Council from 46 in 1976 to 93 in 1979 was a consequence of the 1978 Act which followed the Merrison enquiry into the regulation of the profession and which culminated in a report published in 1975 (HMSO 1975). This enquiry had been set up in response to a rebellion among rank and file doctors in which the legitimacy of the Council to regulate the profession had been called into question. The precipitating issue was a financial one. Until 1969 medical practitioners had on admission to the register paid one lifetime fee. Inflation and increased work led the Council to propose an annual retention fee. Rebellious doctors insisted that if they were to pay an annual fee they wanted a majority of elected, as opposed to nominated or appointed, medical practitioners on the Council.

Changes in the composition of the Council have thus led to an increase in the number and proportion of elected practitioners, with consequent proportionate decreases in those appointed by Royal Colleges, Faculties and Universities. Table 1 also shows that the number of medical practitioners nominated by the Privy Council declined proportionately and absolutely. The two now nominated are the Chief Medical Officer for England and one in rotation of the Chief Medical Officers of Scotland, Wales and Northern Ireland. The British Medical Associa-

Table 1: Council Membership in 1976, 1979, 1984 and 1987

Appointed by	1976	1979	1984	1987
Universities	18	21	21	
Royal Colleges[1]	9	13	13	
Total appointed	27	34	34	34

Elected by registered medical practitioners in				
England & Wales[2]	8			
England[2]		39	39	
Scotland	2	6	6	
Wales		3	3	
Ireland	1			
Northern Ireland		2	2	
Total elected	11	50	50	50

Nominated by Privy Council				
Medical	5	2	2	2
Non-medical	3	7[3]	9[3]	11[4]
Total nominated	8	9	11	13

Total Members	46	93	95	97

(1) Royal Colleges include Faculties
(2) England includes the Channel Islands and the Isle of Man
(3) Includes one nurse
(4) Includes two nurses

Sources: GMC Minutes and Annual Reports, 1976, 1979, 1984, 1987.

tion (BMA, the largest of the medical trade unions and not to be confused with the statutory GMC) is opposed to the presence of any of these civil servants, but the state insists upon the minimum of two. The elections of 1979, by single transferable vote, returned to the Council five "overseas" doctors (who had qualified abroad, mostly from the new Commonwealth; there were none before) and five women doctors, where there had previously been only one.

MY RESEARCH ON THE GMC

My interest in and knowledge about the GMC derives from nine years of service on the Council as a lay member appointed by the Queen in Privy Council. My systematic research on the GMC has been undertaken with a grant from the Economic and Social Research Council (No. G00 232247). The central theme of my analysis is the question of how the GMC performs its difficult task of retaining the confidence and unity of the profession at the same time as it discharges its responsibility to the public (Stacey 1984). This tension is central to the Council's work and is taken seriously by members in authority.

In terms of method, the research is somewhat unusual in that I was a participant before I was an observer. The implications of this will be addressed in the final research publication and have already been discussed in a working paper (Stacey 1986). While a member I took a stance on a number of issues, including some that can be defined as ethical. Thus, for example, I contributed a paper (Stacey 1985) to one of the annual conferences the GMC holds to discuss selected professional matters. That particular conference was devoted to medical ethics.

My paper addressed aspects of the central tension in the GMC's work which was later to become the major theme of my research, for that theme emerges sharply in the domain of medical ethics (Stacey 1985). Historically, the tension dates back to the Hippocratic oath which from its earliest forms demonstrates the two sides of medical responsibility (Jacob 1982). The one is the undertaking to the patient:

> And whatever I see or hear when attending the sick or even apart therefrom, which ought not to be told, I will never divulge but hold as a secret.

The second relates to confidence among practitioners:

> I will teach ... to my sons [sic], the sons of my teacher and to pupils who have sworn the oath of a physician, but to no one else.

This initial advice was directed to a one-to-one doctor-patient relationship. Nowadays many individuals are involved in the treatment of a patient and in the associated ethical issues. Thus, in my conference paper (Stacey 1985: 17) I argued that medical practitioners require greater awareness of the way in which the social organization of the medical profession may lead doctors to emphasize those facets of medical ethics which relate to the practice of the profession rather than those designed to protect the public. I argued also that practitioners need to understand more about the nature of large scale organization if they are to avoid ethical problems which may arise inadvertently simply from the way those organizations work. In addition, I drew attention to the social distance between doctor and patient, the latter's relative powerlessness which leads to the ever present hazard that s/he will be unconsciously treated as an object, rather than as a full and equal human being. In conclusion I suggested that "Council might wish to continue to work to reduce the large grey area in which doctors are not prepared to comment on each others' actions in practice." I pointed out that "a doctor who is incompetent in practice is seen by the public quite simply as 'doing wrong.'" Lay people have no difficulty in seeing incompetence as an ethical issue, I suggested. I could, however, well appreciate the difficulties doctors had in passing judgement on each others' practices. This sociological analysis of medical ethics and its practical implications informs what follows.

THE CHANGING GUIDELINES

Among the recommendations of the Merrison Report of 1975 was that the GMC should offer more positive guidance to practitioners. To do this, the Council had first of all to clear its bona fides with other medical organizations so that this more active role might be accepted. Council set up a Special Committee which held discussions with the BMA and the Defence Societies, i.e. the organizations who insure doctors and to one of which doctors are required to subscribe (GMC 1975: 15; Richardson 1976: 4). These bodies also issue guidance to doctors on ethical and allied matters. Their agreement achieved, work on revisions began. The first revisions were published in 1976; since then several revised guides have appeared, the latest to be discussed here in 1987.

Even at first glance one notices marked differences in the blue pamphlet of 1987 compared with that of 1976. To start with the 1987 version is two and a half times longer (31 pages compared with 18, and much smaller print). The blue pamphlet of 1976 was divided into three sections: Part I described the statutory provisions, the machinery of the Council to do with discipline and something about the way that works; Part II discussed convictions and forms of professional misconduct which may lead to disciplinary proceedings; Part III offered supplementary guidance on specific areas of professional conduct. Some of the changes in the later pamphlet are adjustments to successive administrative reforms. Thus in 1987 a fourth part was added which spoke about the Health Committee established after the 1978 Act.

The contents of the pamphlets are shown in greater detail in the Appendix; it indicates clearly the major changes between the blue pamphlet of 1976 and 1987. Here I will draw attention to some of the more important. The paragraph "Conditional Registration" in Part I is included in 1987 to explain the new powers which the Council now has to make a doctor's continued registration subject to conditions the committee may determine. This constitutes a material difference in the Conduct Committee's disposal options in disciplinary cases. The addition of the phrase "Letters of Advice" to the section on "Warning Letters" not only reflects changes in the procedures but is a response to professional anxieties about the use of the harsh word "warning." In 1987 more information is included about how Council makes decisions to issue such letters.

Part II is designed to give practitioners examples of the sorts of conduct which might get them into difficulties. As the Appendix makes plain, these cover a range of possible offences from drug and alcohol abuse through sexual offences and abuse of patient confidences to offences against professional colleagues. In the 1987 version the intent remains the same but the examples have been grouped into four main headings which distinguish the possible offences differently (see

Appendix). Their purpose seems plainer and the importance of good standards of service to the patient is more clearly stated. Thus discussion of responsibility for standards of medical care now comes first, followed by the issue of abuse of professional privileges which includes a distinction between those conferred by law and by custom.

Part III in 1976 dealt exclusively with advertizing (see Appendix) and was couched in intra-professional terms. In 1987 it has become something really rather different, reflecting the powers given to the Council by the Medical Act of 1983 to provide advice for members of the medical profession on standards of professional conduct or on medical ethics, powers which the Council did not enjoy before 1978. Now more than a supplement to the examples in Part II, the text begins to discuss the philosophy behind the guidance. The section on financial relationships between doctors and independent organizations providing clinical, diagnostic or medical advisory services has been expanded. It has also been turned around so that the best interests of the patient (in the case of a referral to an organization in which a doctor has a financial interest) now comes first, whereas formerly, this guidance appeared as a purely intra-professional matter. An important innovation is the extensive section on relationships between the medical profession and the pharmaceutical and allied industries with subsections relating to proper practice in the clinical trials of drugs, gifts and loans and the acceptance of hospitality.

Part IV, new in 1987, describes the procedures of the Health Committee which only came into being after the 1978 Act. This is the committee to which doctors who are thought to be unfit to practise by reason of physical or mental illness may be referred if they do not voluntarily accept examination and treatment.

HOW THESE CHANGES CAME ABOUT
AND WHAT THEY AMOUNTED TO

This section will present a chronological review of the major ethical preoccupations of the Council between 1976 and 1987 and the changes in guidance which resulted. The origins of the changes will be assessed in the final discussion section.

Advertizing had always been a major ethical concern of the Council. It was the first topic to be revised for the 1976 guidelines which contained new sections in Part III

dealing with questions of advertizing arising from relationships between doctors and organizations providing clinical, diagnostic, or medical advisory services, and questions arising from articles or books, and broadcasting or television appearances by doctors. Another section discusses signposts or noticeboards relating to health centres, and the choice of titles for such centres or for group practice (Richardson 1976: 4).

The organizations mentioned include family planning and vasectomy clinics, pregnancy advisory bureaus and nursing homes including those providing facilities for the termination of pregnancy. All of these are outside the National Health Service (NHS) and thus not subject to NHS disciplinary controls. While some are provident organizations, others are for profit. Many are run by lay persons and advertize either to the lay public or the medical profession. The stress of the guidance has to do with doctors taking unfair advantage of their colleagues either by advertizing their services and skills or by depreciation of other doctors – principally intra-professional issues.

In the following year the Council discussed sexual relations between doctors and patients. According to the President, consensus was difficult to achieve (Richardson 1977: 3). However, in August 1977 new advice was included in the blue pamphlet, taking some account of changing sexual morality in which greater freedom of sexual relations had become more commonly accepted. The tenor of this movement can be seen more clearly in the 1979 edition. Here examples relating to personal relations between doctors and patients and to professional confidence were incorporated in Part II (GMC 1979b: 11) and amplified in Part III (1979b: 14-16; 1979a: 220-221). Attention was drawn to the privileged access to their homes which patients by custom confer on doctors. The need to maintain trust was stressed and three areas where this trust might be breached were pointed to: the improper disclosure of confidential information obtained from or about a patient; improper influence to obtain money from a patient; entering into an emotional or sexual relation with a patient (or a member of the patient's family) "which disrupts that patient's family life or otherwise damages, or causes distress to, the patient or his or her family." There was a clear move away from the simple and uncompromising reference of 1976 to the inappropriateness of adultery (GMC 1976: 9).

The GMC's Annual Report sometimes puts a bit more flesh on the bones of the guidance. Ethical issues, which are not the subject of formal guidelines but which Council feels need reinforcing are raised in it. Thus a section of the 1979 Report included: data about how many complaints were received by Council; firm advice that it was improper for a practitioner to expect a fee from a consultant for referring cases to him; and mention of anxiety, expressed by both public and profession, about advertizements for cosmetic surgery and hair transplant clinics. Council pointed out that it has no jurisdiction over non-medical persons who own or direct such clinics and referred, so far as medical practitioners are concerned, to the existing guidance in the blue pamphlet. The report indicated that the Council was considering whether it should do more (GMC 1980: 24-5).

Newly elected as President in 1980, Sir Robert Wright explained his view of the GMC's role.

Proper professional conduct is based on an understanding and acceptance by doctors of good manners in professional relationships directed to the protection of mutual trust between doctor and patient – medical ethics. Public and professional attitudes change with changing social and economic conditions and with changing external pressures, e.g. advances in technology and exposure to the media for public information. Changes in the availability of potent drugs and in their use and abuse, changes in the distribution of special professional skills, and changes in public concepts of morality can all contrive to produce new challenges and dangers for the medical profession and the public. The Council has a duty to advise the former and protect the latter (Wright 1981: 5).

The work of the Standards Committee noted in the Annual Report for 1980 (GMC 1981a: 17-18) included consideration of confidentiality in relation to computer storage of medical information (the government was also considering data protection at this time), and the issue of advertizing already mentioned. In addition advice had been given to individual doctors by or on behalf of the President that in the case of the reference of a patient to an acupuncturist not registered with the Council, the referring doctor retains ultimate responsibility for the patient; about how doctors might use their names in broadcasts or on television without it becoming improper advertizing; and for the same reasons, about exercising caution when involved with representatives of the non-medical press (GMC 1981a: 18-19).

These deliberations and actions were reflected in a revised edition of the blue pamphlet issued in 1981 which included the following additions: (i) doctors who undertake clinical work for organizations which advertize clinical, diagnostic or medical advisory services to the lay public have a duty to satisfy themselves that the organizations conform to principles acceptable to the profession (GMC 1981b: 19-21); (ii) referrals to specialists should normally come through a general practitioner, and if a patient should be accepted directly by a specialist that person should inform the patient's GP as soon as possible (GMC 1981b: 18-19).

Reference was made to these additions to the blue pamphlet in the Annual Report for 1982 (GMC 1982: 14). That year the Standards Committee also reported that it had considered research on human subjects and recommended that the ethical committees of other medical bodies should be consulted. It reported that it was in the process of examining the question of the prescribing of opioid drugs to addicts by doctors in private practice (GMC 1982: 14-15).

In consequence of further work done by the Standards Committee in 1982, a new edition of the blue pamphlet was issued to doctors in 1983 (GMC 1983; Happel 1983). It was at this time that revised and expanded guidance about dishonesty in relation to improper financial transactions was introduced into Part III (GMC 1983: 13-14). In this edition as well, considerable revisions and expansions were made to the advice about the need for a patient to be able to trust the confidentiality of his or her relationship with doctors (GMC 1983: 18-21). These revisions arose from disciplinary cases (Happel 1983: 39).

The new guidance is important in so far as it spells out questions to do with the sharing of confidential information with non-medical persons caring for the patient (nurses and other health care professionals). The guidance should apply to medical information which a doctor receives in the course of non-clinical as well as clinical duties. Here for the first time, so far as I am aware, Council guidance is really beginning to take account of the organization of modern health care, the many relationships involved and the inevitable responsibilities of others, rather than attempting, as hitherto, to retain the fiction of the one-to-one doctor-patient relation.

The guidance also recommended that doctors, although they should try to persuade minors to confide in their parents, should respect the confidentiality of minors who come to them for contraceptive advice and who do not wish that their parents should be told. This particular piece of guidance was linked to another guidance issued by the Department of Health and Social Security (DHSS as it then was, now the Department of Health) to health authorities and which became the focus of a long drawn out law suit – of which more anon.

Most significant, however, were amendments to the guidance on responsibility for standards of medical care. In Part II the following guidance was included.

> The Council is not ordinarily concerned with errors in diagnosis or treatment, or with the kind of matters which give rise to action in the civil courts for negligence, unless the doctor's conduct in the case has involved such a disregard of his [sic] professional responsibility to his patients or such a neglect of his professional duties as to raise a question of serious professional misconduct. A question of serious professional misconduct may also arise from a complaint or information about the conduct of a doctor which suggests that he has endangered the welfare of patients by persisting in independent practice of a branch of medicine in which he does not have appropriate knowledge and skill and has not acquired the experience which is necessary (GMC 1983: 10).

This inclusion resulted from a case of alleged malpractice which was exposed relentlessly, to the anger of medical members of the GMC, on television and subsequently by the television personality herself in discussions with senior members and officers of the GMC.

Further new advice in part III of this edition relates to relationships between the medical profession and the pharmaceutical and allied industries (GMC 1983: 26-28; Happel 1983: 40). This advice is about accepting payments for participation in the clinical trials of drugs, as well as accepting gifts, loans, and excessively lavish hospitality. Once again, although members sometimes expressed anxiety about the activities of the pharmaceutical industry, this inclusion was not made until after allegations had been made on television about particularly extravagant advertizing of a new drug by a pharmaceutical company which offered doctors a trip on the Orient Express to Venice. The activities thus revealed drew adverse comment in the *Lancet* (1983: 219) and caused the Chief Medical Officer for England to address a public meeting of Council on its responsibilities in such matters.

Many of the major differences noted earlier between the blue pamphlet of 1976 and that of 1987 were introduced in the 1983 edition. This was also the year when a statement was made about female circumcision. In 1982 the DHSS had drawn the attention of the GMC to a report that doctors were circumcising women in the UK. After debate, Council issued a statement which included the following sentence. "The Council regards performance of such an operation in the United Kingdom on other than medical grounds as unethical" (Happel 1983: 41).

A Mrs. Gillick challenged the right of medical practitioners to withhold from the parents of young women who were minors at law the information that their daughters had been given contraceptive advice (Mrs. Victoria Gillick vs. The West Norfolk and Wisbech Health Authority and the Department of Health and Social Security). Mrs. Gillick was encouraged by organizations which sought nation-wide support for her case. A finding in her favour by the Court of Appeal forced the Council to modify the guidance which it had issued in 1983 (GMC 1985a: 19-20, 1985b: 1-2). In the House of Lords, where the Law Lords sit as the final court of appeal, Mrs. Gillick lost (17 Oct. 1985). The Law Lords concluded that the ultimate decision lay with the doctors who should however take parental right into account (GMC 1986: 3). By 1987 blue pamphlet advice on this issue was not only much longer and more circumlocutory, stressing throughout the child's [sic] best medical interests, but also indicating that in some circumstances (for example, where the "child" is thought to be insufficiently mature) the parents may be told against the patient's wishes. However in these cases the patient should be informed of what the doctor proposes to do (GMC 1987: 21).

Changes consequent upon the Gillick ruling were only one of five amendments to the guidance on confidentiality which were made in the 1985 edition of the blue pamphlet (GMC 1985b: 19-20). Following a government circular relating to the use of deputizing services, which general practitioners employ to relieve themselves of house calls, especially at night, the guidance about delegation of medical responsibility was further revised (GMC 1985a: 19-20). This was a matter of public anxiety about the over-use and inadequacy of night-time deputizing services, rather than of intra-professional relations.

Non-medical Members of Parliament had shown little interest in the GMC throughout the debate on the 1978 Act (Stacey 1989). However, one Member, Nigel Spearing, became concerned that a constituent's child had died in consequence of inadequate medical attention and yet a finding of serious professional misconduct could not be brought. He consequently sought in 1984 to bring a private member's Bill which would make it possible to have (as the New Zealand GMC does) a lesser offence of "unacceptable professional conduct." The GMC at its November 1984 meeting rejected that idea. (GMC 1984b: 4-5, 22-23, Appendices VII and XIX; GMC 1985a: 28-29). It decided however to issue more detailed guidance on standards of professional care (GMC 1985a: 28-29). These were incorporated in the blue pamphlet and particular attention was drawn to them in the Annual Report which goes out to all registered medical practitioners (Irvine 1986: 5). Stating that "the public are entitled to expect that a registered medical practitioner will afford and maintain a good standard of professional care," attention is drawn to five points: conscientious history taking, thorough examination and diagnosis, competent and considerate professional management, prompt action when urgent intervention is indicated, readiness to consult colleagues as necessary (GMC 1987: 10).

All in all the blue pamphlet of 1987 is much different from its predecessor of 1976. It still devotes a good deal of space to intra-professional ethics, but there is more in it that is patient-oriented. Specifically a number of issues to do with the confidentiality of the information which a doctor may have acquired about a patient have been developed, as have issues to do with standards of medical care of patients. The extensive section on relations with the pharmaceutical industry should also be mentioned in this connection. How one may explain these changes is the first topic to be considered in the next section.

DISCUSSION

Obviously in one sense all changes in the ethical guidance which is given to doctors come from within the Council. This is the case whether the guidance is printed in the blue pamphlet, included in the Annual Report which goes to all registered practitioners, or, more rarely, is issued in public statements. In all these cases the President and members have become convinced that change is necessary. The question is what led them to come to these collective conclusions? Did they develop spontaneously from discussion which took place in the Council as a result of the experience of work there? Was the question raised in Council by an individual member following experience outside, or in response to outsiders' comments on Council actions? If so, did that arise from her/his own experience or from the views or experiences of another organization to which s/he belongs? Were the problems represented to the President or a member by an individual or an organization outside the Council?

There are, of course, two senses in which pressure for change may originate outside the Council: one is from within the profession but not within the Council itself; the other is from outside the medical profession altogether. In the first sense pressure may originate in other medical organizations, for example, Royal Colleges or Faculties, or in medical trade unions such as the BMA or other medical interest groups. Sometimes these influences can be seen clearly, when, for example, there is an articulated BMA policy developed on an issue or much more rarely when Presidents of the Royal Colleges write formally to the President of the GMC. In other cases the origin of the influence may be more vague, as when opinion has begun to coalesce round a topic in medical common rooms here or there.

The second kind of pressure is lay. It may be expressed in the media, as we saw in one case mentioned above; by Members of Parliament to Ministers in the House; it may arise from within the Department of Health. Pressure from the DHSS may, of course come from medical administrators, but whether this can be called medical in a strict sense is doubtful, for these may be conveying messages from Ministers or from the House. It is in part opposition to pressure of this sort that leads some doctors to feel that Chief Medical Officers should not be full members of the Council.

The issue of confidentiality in the case of minors is an interesting example of the way outside pressures can work. Many parents, some but not all of whom conscientiously object to contraception, were disturbed that their daughters could get contraceptive advice without their knowledge. Initially, Council in 1983 issued a guidance to the effect that the doctor's responsibility to retain confidentiality must be maintained where a patient refuses to allow a parent to be told. Some months later, it decided to add to this the phrase, "whether or not he [the doctor] decides to offer advice or treatment" (GMC 1984: 16). This occurred because some doctors had indicated that the original wording could be seen as implying that a doctor would automatically give contraceptive advice to a minor who asked for it.

Disquiet on the part of some members of the public – Mrs. Gillick and her supporters – was upheld by the law at some levels but not in the House of Lords, the final Court of Appeal. Consequently the GMC throughout the hearings had to alter its guidance to keep in line with the latest move in the developing case law.

While undoubtedly there were dissenting doctors, most medical practitioners felt that the hazards of early teenage pregnancies, coupled with their general duty to maintain the confidentiality of the patient, impelled them both to offer contraceptive advice and to respect the confidentiality of those young women who did not wish their parents to know they had consulted about contraception. In this case the interests of one section of the public, the young patients, went along with the interests of the medical practitioners whose autonomy to act as they thought best, so long as they took account of parental interests, was sustained by the judgement at the final appeal. This perspective of medical practitioners, and their concepts of professional autonomy which lawyers share, may have informed the outcome as much as any legally recognized rights of young women to control their own fertility. Or, to put it another way, had there not been this coincidence the outcome may have been different.

The ground swell among medical practitioners which led to the Merrison Inquiry and the 1978 Act has been mentioned. A number of the resultant changes are further examples of the first type of outside pressure: that which comes from within the profession but outside the Council, e.g. professional anxieties about the Council's disciplinary functions, the way these were exercised and their effect upon the practitioners involved. These concerns seem to have influenced the selection of the first topics for the new guidance. For example, I noted above that the guidance relating to family planning and vasectomy clinics and the like, many of which were privately run, stressed concerns to do with relations between professionals rather than the doctor's responsibility to see that the information made available to the public is not misleading. In this sense all these new additions appear to be of more importance to the profession than the public.

It was only in 1983, when a section on standards of care was introduced, that any clearly discernable move towards assuring competence to practise and protecting the public took place. From its foundations in 1858 and until well into the mid-twentieth century, the Council had much of the appearance of a London gentleman's club, concerned only that its members behave like gentlemen and that "bounders" should be evicted. Changes in the guidance represent a continuing attempt on the part of medical leaders to transform the Council into a body which demonstrates that the medical profession can handle the privileges of professional self-regulation in the late twentieth century. However, it is one thing to make the guidelines, it is another to follow them through rigorously in the disciplinary process. The President and other senior Council members may have recognized the need to keep at least one step ahead of public opinion expressed through government channels, but it may have taken time for others to recognize and fully accept the need for the changes made.

Implementing the new guidelines was up to the practitioners sitting on the various committees associated with discipline and conduct. Table 2 indicates the number of cases of alleged or serious professional misconduct involving such issues as a failure to visit when reasonably requested or to treat or diagnose properly or to refer as necessary, in relation to the total number of cases brought. New cases coming forward to the Disciplinary Committee in the period 1976-1982 and to the Professional Conduct Committee in 1980-1984 are distinguished. (From 1980 Professional Conduct took over the work of the Disciplinary Committee; the overlap in dates results from the winding down of the work of the latter.) Cases relating to abuse of drugs or alcohol have been excluded since after the 1978 Act the majority of these were sent to the Health Committee.

As Table 2 shows there was a slight increase in the proportion of charges which could be said to be related to competence to practise or competence in practice: ten out of 51 compared with 34 out of 130. What the table cannot show is that the proportion was increasing throughout the period, almost three quarters of the cases coming in 1983 and after. This suggests that in the latter period more such offences were gradually being sent forward from the Preliminary Proceedings Committee. However, reading the cases in detail, one has the impression that they were not dealt with as strictly as one might expect. How far do the data bear out this impression?

The Table shows that 8 out of 10 individuals charged with failure to visit or care for cases were found guilty in the first period compared with 23 out of 130 in the second. None was erased in the first period and 3 in the second. Taking all cases heard in the first period, excluding alcohol and drug offences there were 5 erasures (for sexual offences, 1; fraud, 1; offences against colleagues, 3). In the second period there were 16 erasures in all (including: for sexual offences, 7; fraud, 1; offences against the profession, 5).

It would appear that not only were more cases related to competence to practise coming through, but that they were being taken more seriously than heretofore. The impression that they were not dealt with as strictly as others arises in part from the nature of the eight cases which, while receiving a finding of guilty, were dismissed with admonishment only. All these cases went through

Table 2: Allegations of serious professional misconduct on account of failure to visit, treat, prescribe, refer appropriately and of maltreatment and incompetence and the outcomes.

| | New Cases | |
| | 1976-1982* | 1980-1984* |
Alleged Offence	Disciplinary Committee	Professional Conduct Committee
	Cases	Cases
Failure to visit, treat etc	10	34
Cases found guilty	8	23
Of these: erased	0	3
suspended	2	3
registration made conditional	NA	1
admonished	2	8
concluded	1	1
to Health Cttee	NA	2
postponed decision	2	4
adjourned	1	1
Total cases in period**	51	13

* There is an overlap of two years between the beginning of the Professional Conduct Committee and the ending of the Disciplinary Committee.

** To improve comparability, new cases involving drugs and alcohol have been excluded from this table; the establishment of the Health Committee at the same time as the Professional Conduct Committee sharply reduced the number of those cases which were treated as disciplinary offences, the majority after the 1978 Act being considered as health matters, whereas all had previously come to the Disciplinary Committee.

NA = not applicable

Source: GMC Minutes 1976-1984

three preliminary stages before reaching the Conduct Committee. Furthermore, the proportion of complaints the Council received which eventuated in disciplinary hearings was very small compared with the total. In 1974 of 847 received, less than 100 went so far as the Penal Cases Committee, while the Disciplinary Committee dealt with 29 (about 3%) in that year (GMC 1975). From September 1985 to August 1986, 798 complaints were received of which 167 went to the Preliminary Proceedings Committee which passed 49 (6%) on to the Professional Conduct Committee (GMC 1987: 15-16). One may reasonably assume that the small proportion of cases which reach the Professional Conduct Committee are among the more flagrant, having come through this entire procedure.

Polls and surveys show that the public at large is well satisfied with the service it receives in the NHS. Nevertheless there is always a small proportion of complaints, few of which are formerly put to any authority. The number reaching the GMC is a minute proportion of these few. Among bodies to which members of the public complain, the GMC probably seems the most remote. Many do not complain to it because it seems difficult of access. Nor does the Council offer its services to the public. That is not how it defines its task. The blue pamphlet on professional conduct, fitness to practise and medical ethics is addressed to medical practitioners. It is instructive to imagine how differently it might be written if the authors were addressing the general public. The GMC publishes no pamphlet which specifically spells out these issues for the public so that patients might know what reasonably to expect from doctors in the way of professional behaviour and what are the procedures in cases of alleged violation. The relationship of GMC with NHS procedures is by no means transparent – although I hear that discussions are now going on with a variety of organizations as to how this problem might be overcome.

In this sense professional self-regulation does appear in the past at least to have been a somewhat inward-looking matter in which doctors themselves defined what was unprofessional conduct and decided when other doctors were not behaving properly. It is not, however, a matter of organizational arrangement alone, although changes in the organization of the disciplinary procedures could have important consequences. There are differences in perception of practice between patients and practitioners which suggest that the problem is more deeply rooted than organizational changes alone could overcome. The difficulties lie in issues to do with the kind of knowledge and understanding the doctors and patients have about the treatment process, about matters that may go wrong and about the factors which each take into account in decision-making. I would like to illustrate these differences – differences which go some way to explaining what seems to be a bias towards practitioners and away from the public – through the findings of a recent empirical study of the ethics of informed consent to cardiac surgery on behalf of children.

Priscilla Alderson (1988), in a sensitive study of the problems of parents' informed consent to paediatric cardiac surgery, has shown how difficult the decisions are which medical practitioners have to make when considering whether to propose the operation; how difficult also are the decisions that the parents face when surgery is suggested. She shows that the nature of the decisions and what is taken into account differ depending on whether one is on the parental or the medical side of the child's bed, so to speak. The nature of the knowledge that each uses, the conspectus of what it relates to, what social context is taken into account, are all shown to be quite distinct.

While for the medical team the decision to propose an operation takes place at a particular time, there is no one moment when parents can give informed consent. The child and parents are drawn ineluctably onto a course from which it becomes increasingly hard to escape. Structured and systematic differences separate the parental and the medical care giver, making the goal of mutuality in the child's interest that both want to achieve remarkably hard to realise.

The understanding of the cardiac physicians and surgeons derives from their general medical knowledge, their specialist cardiac knowledge and their judgement about the particular case. The parents' knowledge is singular, acquired through the experience of having a child who is a cardiac patient and of observing the experiences of similarly afflicted children and their parents. Their knowledge takes account of very much more than the state of the child's heart (although that is of prime importance to them) and takes account in a different mode. Their knowledge includes, for example, their understanding of what it might be to have to care for a child who is alive but brain damaged, both conditions a possible consequence of the operation.

In consequence of the different sorts of knowledge which medical practitioner and parent are using to assess the situation, it is not possible for them to communicate fully. The parents may be ignorant of many details of which the cardiac team are thoroughly appraised. However, the medical practitioners remain at least as ignorant of what the patient's parents know. In this ignorance, it is hard for medical practitioners to come to a fully ethical decision involving as well meaningfully informed consent. The crucial point is that each is working with a different kind of knowledge.

These are the sorts of differences which lead patients and practitioners to judge differently, sometimes radically so, as to whether the Council fulfills its responsibilities to the public; they also lead to lay feelings that doctors are insufficiently tough on their incompetent colleagues. Furthermore, the ethics of decisions of the kind which Alderson's medical practitioners have to take, and which her evidence shows they take carefully and seriously, seem not to be touched by the sort of guidance which appears in the blue pamphlet. That document and the Council itself deal with ethics of a different kind. Happily the particularly acute ethical dilemmas which Alderson describes do not arise for the majority of parents or of medical practitioners. Their acute form, however, serves to highlight what are much more general problems in the handling of medical ethics and helps to reveal the underlying patterns.

Differences such as these about what is reasonable professional behaviour assail the attempts of the General Medical Council to regulate the profession in a way which is plausible or salient for the generality of the lay public. It is not only that the material interests of the public and the profession may diverge, as indeed they may, it is also that different problems and, most importantly, different ways of understanding the problems, are involved. Since the profession is self-regulating and the lay voice – even the enlarged lay voice – is small, it is not surprising if, in the tension between maintaining the unity of the profession and protecting the public, the latter seems, in the past at least, to have come off second best. Whether the moves that the Council has made in recent years towards issues of greater interest to patients will enable it to strike a better balance in the future remains to be seen.

ACKNOWLEDGEMENTS

The first draft of the paper was written in the Department of Social and Behavioral Sciences, School of Nursing, University of California, San Francisco, when I held the 1988 Lucile Petry Leone Distinguished Visiting Professorship. I gratefully acknowledge the facilities and the support made available to me by the Department, the School and the University. I also wish to acknowledge the help of members and officers of the Council in revising the paper. My gratitude also to Judy Morris and Eileen Clark for searches and analyses.

APPENDIX

Contents of the "Blue Pamphlet": 1976 and 1987 compared.

1976

Part I

Statutory Provisions; Convictions
The Meaning of "Serious Professional Misconduct"
The Disciplinary Committee and the Penal Cases Committee; Rules of Procedure;
   The Earlier Stages of Proceedings
Warning Letters; Inquiries before the Disciplinary Committee
Powers of the Disciplinary Committee at the Conclusion of an Inquiry;
   Postponement of Judgment
Suspension of Registration; Erasure; Appeal Procedure and Immediate Suspension
Restoration to the Register after Disciplinary Erasure

Part II: Convictions and Forms of Professional Misconduct which May Lead to Disciplinary Proceedings

(i)   Disregard of personal responsibilities to patients
(ii)   Abuse of alcohol
(iii)   Abuse of controlled drugs
(iv)   Termination of pregnancy
(v)   Abuse of professional position in order to further an improper association or
          commit adultery
(vi)   Abuse of professional confidence
(vii)   Offences involving dishonesty, indecency or violence
(viii) Advertising: Depreciation of other doctors
(ix)   Canvassing and related offences
(x)   Untrue or misleading certificates and other professional documents
(xi)   Improper delegation of medical duties to unregistered persons and covering
(xii)   Improper financial transactions

Part III: Supplementary Guidance in Specific Areas of Professional Conduct

(i)   Questions of advertising arising from relationships between doctors and
          organizations providing clinical, diagnostic, or medical advisory services
(ii)   Questions of advertising arising from articles or books, broadcasting or
          television appearances by doctors
(iii)   Signboards or notice boards relating to health centres or medical centres

Conclusion

Source: GMC 1976.

1987

Part I: The Disciplinary Processes of the Council

Statutory Provisions
Convictions
The meaning of "serious professional misconduct"
The Professional Conduct Committee and the Preliminary Proceedings Committee
Rules of procedure
Proceedings: the preliminary stages
Powers of the Preliminary Proceedings Committee: warning letters and letters of advice
Inquiries before the Professional Conduct Committee
Powers of the Professional Conduct Committee at the conclusion of an inquiry
Postponement of determination
Conditional registration
Suspension of registration
Erasure
Appeal Procedure and immediate suspension
Restoration to the Register after disciplinary erasure

Part II: Convictions and Forms of Serious Professional Misconduct which May Lead to Disciplinary Proceedings

Neglect or disregard of professional responsibilities to patients for their care and treatment
    Responsibility for standards of medical care
    Delegation of medical duties to professional colleagues
    Delegation of medical duties to nurses and others
Abuse of professional privileges or skills
    Abuse of privileges conferred by law: Misuse of professional skills
    – Prescribing of drugs
    – Medical certificates
    – Termination of Pregnancy
    Abuse of privileges conferred by custom: Professional confidence; Undue influence;
        Personal relationships between doctors and clients
Personal Behaviour: conduct derogatory to the reputation of the profession
    Personal misuse or abuse of alcohol or other drugs
    Dishonesty: improper financial transactions
    Indecency and violence
Self-promotion, canvassing and related professional offences
    The use of promotional material
    Canvassing and other improper arrangements to extend a doctor's practice
    Disparagement of professional colleagues

Conclusion: The nature of serious professional misconduct

Part III: Advice on Standards of Professional Conduct and on Medical Ethics

Personal relationships between doctors and patients
Professional confidence
Principles governing the reference of patients to, and their acceptance by, doctors
    providing specialist services
  Reference of patients to specialists
  Acceptance of patients by specialists
Self-promotion: circumstances in which difficulties most commonly arise
  Notices or announcements by or about doctors
  Relationships between doctors and organisations providing clinical, diagnostic
    or advisory services
  Public references to doctors by other companies or organisations
  Articles, books and broadcasting by doctors
Financial relationships between doctors and independent organisations providing
    clinical, diagnostic or medical advisory services
Relationships between the medical profession and the pharmaceutical and
    allied industries
  Clinical trials of drugs
  Gifts and loans
  Acceptance of hospitality

Part IV: Fitness to Practise: Procedures Associated with the Health Committee

Source: GMC 1987

REFERENCES

Alderson, D.P.
    1988   Informed Consent: Problems of Parental Consent to Paediatric Cardiac Surgery. Ph.D. Thesis. London University.

GMC (General Medical Council)
    1975   Annual Report for 1974. London: General Medical Council.
    1976   Professional Conduct and Discipline. London: General Medical Council.
    1978   Annual Report for 1977. London: General Medical Council.
    1979a Minutes of the General Medical Council. Vol. CXVI. London: General Medical Council.
    1979b Professional Conduct and Discipline. London: General Medical Council.
    1980   Annual Report for 1979. London: General Medical Council.
    1981a Annual Report for 1980. London: General Medical Council.
    1981b Professional Conduct and Discipline: Fitness to Practise. London: General Medical Council.
    1982   Annual Report for 1981. London: General Medical Council.
    1983   Professional Conduct and Discipline: Fitness to Practise. London: General Medical Council.
    1984a Annual Report for 1983. London: General Medical Council.
    1984b Minutes of the General Medical Council. Vol. CXXVI. London: General Medical Council.
    1985a Annual Report for 1984. London: General Medical Council.
    1985b Minutes of the General Medical Council. Vol. CXXVII. London: General Medical Council.
    1986   Annual Report for 1985. London: General Medical Council.
    1987   Professional Conduct and Discipline: Fitness to Practise. London: General Medical Council.

Happel, J.
    1983   Advice on Standards of Professional Conduct and on Medical Ethics. General Medical Council Annual Report for 1982. London: General Medical Council. Pp. 38–41.

Hill, D.
    1981   The Health Committee Procedures. General Medical Council Annual Report for 1980. London: General Medical Council. Pp. 16–17.

HMSO (Her Majesty's Stationary Office)
    1975   Report of the Committee of Inquiry into the Regulation of the Medical Profession (Merrison), Cmnd 6018. London: HMSO.

Irvine, D.H.
    1986   The Work of the Committee on Standards of Professional Conduct and on Medical Ethics. General Medical Council Annual Report for 1985. London: General Medical Council. Pp. 3-7.

Jacob, J.M.
    1982   Changing Practice on Confidentiality: A Cause for Concern. Commentary I. Confidentiality: The Dangers of Anything Weaker than the Medical Ethic. Journal of Medical Ethics 8: 18-21.

Richardson, J.
    1976   The Council and the Merrison Report. General Medical Council Annual Report for 1975. London: General Medical Council. Pp. 1-5.
    1977   The Council. General Medical Council Annual Report for 1976. London: General Medical Council. Pp. 2-4.
    1980   The Metamorphosis of the Council. General Medical Council Annual Report for 1979. London: General Medical Council. Pp. 4–7.

Stacey, M.
    1984   Application to the ESRC. Coventry: University of Warwick (Mimeo).

1985 Medical Ethics and Medical Practice: A Social Science View. Journal of Medical Ethics 11: 14-18.

1986 From Being a Native to Becoming a Researcher: Meg Stacey and the GMC. Paper presented to the British Sociological Association Medical Sociology Group Conference, York. Coventry: University of Warwick (Mimeo).

1989 The General Medical Council and Professional Accountability. Public Policy and Administration 4: 12–27.

Wright, R.

1981 The GMC: Retrospect and Prospect. Annual Report for 1980. London: General Medical Council. Pp. 4–6.

DAVID J. ROTHMAN

HUMAN EXPERIMENTATION AND THE ORIGINS OF BIOETHICS IN THE UNITED STATES

I

The field of medical ethics has a long history, beginning with Hippocrates and continuing right through the twentieth century. Until the 1970s, however, its most distinguishing characteristic was the extent to which it was dominated by practicing doctors, rather than by philosophers or lawyers. Physicians typically wrote the texts and, no less important, other physicians usually read them. In the mid-1950s, the noted psychiatrist Karl Menninger aptly commented:

> With the one stellar exception of Catholic moralists, there is a strange blindspot about the ethics of health and medicine in almost all ethical literature. Volume after volume in general ethics and in religious treatises on morality will cover almost every conceivable phase of personal and social ethics *except medicine and health* (Foreword to Fletcher 1954: viii-ix).

One of the first contemporary philosophers to break with this tradition, John Fletcher, joined with Menninger to note that "It is a matter for wonder that the philosophers have had so little to say about the physician and his medical arts" (Fletcher 1954: xiv).

Medical ethics as conceptualized and presented by physicians had a very practical bent, concerned not with first principles but with formulating maxims of a do and don't quality. Predictably, too, the formulation of the problems and solutions reflected the vantage point of the doctor – setting forth his rights and responsibilities, how he should behave toward his patients and toward his colleagues.

To a contemporary reader, the ethical tracts of Anglo-American doctors seem most remarkable for elevating medical etiquette over medical ethics, for moving ever so nimbly from the high minded injunction – do no harm – to practical, professionally self-serving maxims – do not slander a fellow doctor or pay social visits to his patients or contradict him in front of his patients. In this literature, coveting a colleague's patients appears as a far more serious breach than coveting his wife. But the distinction between etiquette and ethics hardly mattered to the tracts' authors or audience of practicing physicians. These were not exercises in philosophical discourse but manuals for the man in the field who was equally intent on doing well by his patient and doing well by his practice.

Although etiquette dominated, it would be a mistake to dismiss all the tracts as business manuals, tiresome variations on the theme of how to win patients and influence colleagues. In the United States, for example, Benjamin Rush, the celebrated Philadelphia physician and signer of the Declaration of Independence,

G. Weisz (ed.), Social Science Perspectives on Medical Ethics, 185–200.
© 1990 Kluwer Academic Publishers. Printed in the Netherlands.

delivered a series of lectures on the "vices and virtues" of the physician. The virtuous physician was not pecuniary-minded, and recognized an obligation to provide his services without regard to patients' ability to pay. He was abstemious, even ascetic.

> The nature of his profession renders the theatre, the turf, the chase, and even a convivial table in a sickly season, improper places of relaxation to him.... Many lives have been lost, by the want of punctual and regular attention to the varying symptoms of diseases; but still more have been sacrificed by the criminal preference, which has been given by physicians to ease, convivial company, or public amusements and pursuits, to the care of their patients.

Indeed, Rush's ethical physician was not merely dedicated but heroic: should a plague should strike a community, Rush argued uncompromisingly that the physician was obliged to stay and treat, even at the risk of death (Rush 1947: 293-307).

In this same tradition and a century later, Dr. Richard Cabot, Professor at the Harvard Medical School, both expanded the range of issues that belonged to medical ethics (such as fee splitting) and gave unconventional answers to conventional questions (on birth control and contraception). Cabot, for example, was uncompromising on the need to tell patients the truth (not for him to repeat Holmes' sentiments that a patient was no more entitled to the truth than he was to all the medicines in a doctor's saddle bags). Physician-dominated medical ethics, then, was not just a self-serving, low-minded enterprise. However, it often lacked intellectual rigor and was generally accorded an insignificant place in medical education. "I know no medical school in which professional ethics is now systematically taught," Cabot wrote in 1926, an observation that continued to be true through the 1950s (Cabot 1926: 23).

Just how much medical ethics was – and was expected to be – the doctors' preserve emerges with particular clarity in the pronouncements of the American Medical Association. In the 1930s, for example, the AMA justified its opposition to a variety of economic initiatives (group practice, national health insurance, and physician advertising) by an appeal to a timeless and physician-defined medical ethic. "Medical economics" declared the AMA, "has always rested fundamentally on medical ethics," whose principles, it insisted, did not vary by time (ancient or modern) or by place (Europe or the United States) or by type of government (monarchy or republic). The guiding maxims were constant and "such continuous persistence through so wide diversity of environments seems to prove that judged by the 'survival test,' medical ethics has demonstrated its essential social soundness." This success reflected the fact that the ethic was empirically based, that is: "Each new rule or custom was tested in actual practice," and then judged in terms of how well it supported the "close personal relationship of the sick person and his trained medical adviser," and how well it "promoted the health of patients, as measured by morbidity and mortality statistics.... Ethics thus becomes an integral part of the practice of medicine. Anything that aids in the fight on disease is 'good'; whatever delays recovery or injures health is 'bad.'" With a confidence that tipped over into arrogance, the AMA smugly concluded: "This development and treatment of medical ethics gains much greater significance when compared with the development of ethics in society as a whole." Medical ethics, unlike ethics in general, had been spared "metaphysical or philosophic controversies," such as those which pitted utilitarians against Kantians (AMA n.d.: 58-61).

All this added up to an insistence that medical ethics should be left entirely to medicine, and whatever public policies flowed from these ethical principles were not to be contested or subverted. Any undercutting of the centuries-old precepts was an invitation to social chaos: "Ethical rules and customs are among the most important of the stabilizing elements in society" (AMA n.d.: 63). Thus, because "this close personal relationship of the sick person and his trained medical adviser" was basic to the ethic, government interference with medicine (through health insurance) was to be resisted, as was the establishment of group practices (58). In sum, just as medical ethics was the exclusive preserve of the doctor, the implications of medical ethics served to make all of health policy the preserve of the doctor. Outsiders lacked the moral or intellectual justification to intrude.

Such claims did not go entirely uncontested. There were some non-physicians writing in the field of medical ethics, particularly those coming from a Catholic tradition. Although Jewish and Protestant theologians and philosophers generally ignored medical ethics, Catholics filled publications like the *Linacre Quarterly* with articles applying dogma to medical questions: in the case of a tubal pregnancy, was it permissable to remove the fetus, in effect killing it, in order to save the mother's life, or was it necessary to leave the fetus in place, although the death of the mother as well as the fetus would ensue? The answer focused on direct versus indirect harm – with death to the fetus allowed if it was the indirect result of treatment of the tubal defect (*Linacre Quarterly* 1941: 6-23). The Catholic involvement in medical ethics was actually strong enough to inspire bitter criticism. Paul Blanchard, for example, wrote a blistering attack on the authority that Catholic priests ostensibly exercised over Catholic doctors and nurses in their professional lives. But whatever one thought about the proper limits to church influence, Catholics were not going to leave medical decision-making to the doctors (Blanchard 1949).

Another challenge to the hegemony of physicians over medical ethics came from Joseph Fletcher in his pioneering book, *Morals and Medicine* (1954). Fletcher also entered the field through religious ethics, but in his case, Protestant rather than Catholic ethics. Had Fletcher simply performed another exercise of doctrinal exegesis, relating dogma to one or another medical practice, there would be little reason to single out his book. But *Morals and Medicine* did something novel, purposefully examining medical ethics "from the patient's point of view." Although the subject matter that Fletcher explored was relatively traditional – the ethics of reproduction and euthanasia had a long history in

Catholic literature – his approach was not. Fletcher's guiding premise came from outside medicine: to act as a moral being, an individual had to have the knowledge and freedom to make choices, for otherwise, "we are not responsible; we are not moral agents or personal beings." Bringing this principle to medicine meant that a patient kept ignorant and rendered passive by his doctor was not a moral agent but a puppet, "and there is no moral quality in a Punch and Judy show" (35). Hence, physicians were obligated to tell patients the truth about a diagnosis and condition, not to satisfy a professional medical creed, but because patients had to be allowed to exercise choice.

This same assumption guided Fletcher's positions in the area of reproduction. In vigorous disagreement with Catholic doctrine, he argued that contraception, artificial insemination, and sterilization were not "unnatural acts," but procedures which enhanced individual choice. Contraception, for example, "gives patients a means whereby they may become persons and not merely bodies" (97). Artificial insemination was a method that kept the accidents of nature (sterility) from overruling "human self-determination" (142). Sterilization also was to be left to individual preferences because "moral responsibility requires such choices to be personal decisions rather than natural necessities." Even active euthanasia passed Fletcher's test since the alternative, "to prolong life uselessly, while the personal qualities of freedom, knowledge, self-possession and control, and responsibility are sacrificed is to attack the moral status of a person" (191). Thus, Fletcher moved medical ethics away from the prerogatives of the physicians and from the essentials of a religious creed to the prerogatives of the patient. Nevertheless, he did not immediately stimulate a new approach to medical ethics. For at least another decade, the book remained an odd contribution, not, as it appears in retrospect, the beginnings of a new departure.

To revolutionize the field of medical ethics would require more than one man following a new approach. It would require an entirely new set of questions to come to the fore, questions which would prompt a wider public and a core of physicians to conclude that doctors alone, following the traditional precepts of medical ethics, could not resolve them. And that is precisely what happened in the early 1960s.

II

The first of these new questions, the initial impetus to change, occurred in the area of human experimentation. A series of exposés involving the failure of researchers to obtain subjects' consent to experimentation sparked widespread indignation. The reaction did not focus as much on the character of the investigators themselves as on the nature of the research enterprise. The conclusions drawn from the incidents had less to do with individual malfeasance than with a critique of what might well be called the research ethic. Many outside and some inside the medical community became persuaded that investigators' readiness to make utilitarian judgments, their willingness to sacrifice the interests of some to the interests of others, rendered them incapable of deciding what was or was not ethical in the conduct of human experimentation. The investigator, unlike the physician, could not be trusted to act in the best interests of his subject-patient. And it was from this premise that philosophers, lawyers, and legislators, who had heretofore devoted little attention to medicine, became deeply concerned with research ethics. A new group of commentators and actors had entered the field, first to become involved with medical experimentation, and eventually, with medical therapies.

One useful starting point for tracing the dynamics of this change is Henry Beecher's 1966 article, "Ethics and Clinical Research." As has so often happened in the course of American history, a publication like *Uncle Tom's Cabin*, *The Jungle*, or *Silent Spring* will expose a problem, whether it be the breakup of the black family under slavery or contaminated food or deadly pesticides, in such compelling fashion as to transform public attitudes and policy. Beecher's *New England Journal of Medicine* piece belongs to this tradition. Its indictment of research ethics helped begin the movement that brought a new set of rules and a new set of players to medical decision-making (Rothman 1987).

The article was short, barely six double-columned pages, and the writing terse and technical, primarily aimed at a professional, not a lay, audience. It tried, not altogether successfully, to maintain a tone of detachment, as though this were a scientific paper like any other. "I want to be very sure," Beecher remarked in a letter, "that I have squeezed out of it all emotion, value judgments, and so on."[1] Even so, its publication created a furor, in and out of the medical profession. Beecher was certainly not the first to call attention to the abuses occurring in the conduct of human experimentation and an effort to formulate stricter policies was already underway at the National Institutes of Health. However, his presentation of 22 examples of investigators who endangered "the health or the life of their subjects," without informing them of the risks or obtaining their permission, "made an immediate and vital contribution to U.S. federal policy" in its formative period (Faden and Beauchamp 1986).

At its core were capsule descriptions of the 22 case examples. Example 2 was about the purposeful withholding of penicillin from servicemen with streptococcal infections in order to study alternative means for preventing complications. The men were totally unaware of the fact that they were part of an experiment, let alone at risk of contracting rheumatic fever, which twenty-five of them did. Example 16 involved the feeding of live hepatitis viruses to residents of a state institution for the retarded in order to study the etiology of the disease and attempt to create a protective vaccine against it. In Example 17, physicians injected live cancer cells into twenty-two elderly and senile hospitalized patients without telling them that the cells were cancerous, with a view to examining the body's immunological responses. Example 19 described how researchers inserted a special needle into the left atrium of the heart of subjects, some with cardiac disease, others normal, in order to study the functioning of the heart. In the final example, number 22, researchers inserted catheters into the bladders of 26 newborns less than 48-hours-old and then took a series of x-rays of their bladders filling and voiding as a means of examining the process. "Fortunately," noted Beecher, "no infection followed the catheterization. What the results of the extensive x-ray exposure may be no one can yet say."

Beecher's most significant, and appropriately most controversial, conclusion was that "unethical or questionably ethical procedures are not uncommon" among researchers, that a disregard for the rights of human subjects was widespread. The twenty-two cases, he declared, had been too easy to compile; an earlier and longer draft of the article had a total of 50 which had to be winnowed down for publication. Moreover, the research Beecher was describing was not the work of a few marginal investigators working out of backyard woodsheds. Experiments on servicemen pointed to the cooperation of the U.S. Army; research on the institutionalized retarded implied the approval of state boards of mental hygiene. And research involving cardiac catheterization could only be conducted in a major university or government hospital. In all, Beecher was attacking the research establishment for presuming to decide who should be making what sacrifices in the war against disease, which patients would be the martyrs to science.

The public response to the article was one of anger mixed with disbelief. Feature stories appeared in the leading newspapers and weeklies, which was very much what Beecher had intended. The accounts faithfully reported the details of the experiments, questioning how respectable scientists could have done such things. The official reaction, at places like the National Institutes of Health (NIH), was to move quickly to implement more formal and rigorous procedures governing the conduct of human experimentation. If Beecher's examples represented the state of the research art, it would be foolhardy for anyone to trust to investigators to protect the subjects of their experiments.

The reaction to the exposés of clinical research and human experimentation marked a turning point in the United States in the history of both medical regulation and bioethics. For the first time and in direct response to abuses, decisions that had traditionally been left to the individual conscience of physicians now came under collective surveillance. Federal regulations, a compulsory system of peer review, assurances by universities and hospitals that they were monitoring the research, specific criteria that investigators had to satisfy, and a list of proscribed activities replaced the reliance on the researchers' good will and moral sense. At the same time, the ethics of human experimentation became one of the first issues to bring an extraordinary array of lawyers, philosophers, clergymen, and elected officials into the world of medicine. Thus, the events in and around 1966 accomplished what the Nuremberg trials had not: to bring the ethics of medical experimentation into the public domain and to make apparent the consequences of leaving decisions about clinical research exclusively to the individual investigator.

The new rules were not at the outset – or even when extended and refined over the next two decades – as intrusive as some investigators feared or as protective as some advocates believed necessary. At their core was the operation of an Institutional Review Board (IRB), a committee whose members (mostly but not exclusively drawn from the ranks of medicine) had to pass on the investigator's proposed research involving human experimentation. The effectiveness and authority of the IRB had important limitations, but nevertheless, their most remarkable feature is that a doctor who wore a laboratory coat could no longer unilaterally decide on the ethics of his research. He now had to answer first and formally, to a committee operating under federal guidelines, and second and more informally, to an interested group of observers, many of whom came to carry the label, bioethicist.

The transformation, as might be expected, was not easily accomplished. The tradition of physician discretion was powerful, appealing, and carried ample justifications. Indeed, just how powerful it was can be seen in Henry Beecher's equivocal response to his own findings. Having laid out the dismal record of researchers' disregard for the well-being of their subjects, he was nevertheless reluctant to propose a system of rules or regulations. He was prepared to offer a short list of recommendations, do's and don'ts for investigators: use great caution in seeking volunteers from among laboratory personnel or medical students (for they might feel coerced to consent because of their positions); it was permissible to conduct research on prison inmates provided they gave consent. But he was disinclined to have a government body issue regulations or a professional body, binding codes. Such efforts, he was convinced, would have little impact on researchers' behavior. He did think that a "group decision supported by a proper consultive body" might be useful – and over the 1970s, came to embrace this solution more enthusiastically; but Beecher was mostly persuaded that the investigators in his twenty-two protocols had been more ignorant than malicious. With more of a rhetorical flourish than hard evidence he insisted that their "thoughtlessness and carelessness, not a willful disregard of the patient's rights, account for most of the cases encountered." Accordingly, "conscience-raising," an appeal to professional trust and responsibility, was response enough: "Calling attention ... will help to correct abuses present." For all the abuses he exposed, Beecher still wanted to rely on the integrity of the individual researcher. After weighing all the alternatives, "the more reliable safeguard [is] provided by the presence of an intelligent, informed, conscientious, compassionate, responsible, investigator."

If a Henry Beecher was so hesitant about reining in the researcher, how did any type of governance come about? In the first instance, the exposés of research practices had an enormous impact on the administrative leadership of the research community, particularly on the NIH. Because research support and direction was so highly centralized in the offices of the NIH, the repercussions were immediately felt and acted upon. Thus, the pioneering efforts to create some kind of regulatory mechanisms came from inside the research establishment. Insiders, to be sure, were not inclined to adopt the most far-reaching or rigorous control measures. But their responses ensured that the exposés would not be a simple matter of capturing a headline in today's newspaper and then being forgotten within the week. Instead, the NIH had to evaluate how much political fall-out it would face if it left policies unchanged – and it found the price of inaction higher than the price of action.

NIH had not been caught altogether unawares. The disclosures at the 1962 Kefauver drug hearings that physicians were administering experimental drugs without informing patients prompted James Shannon, then head of the NIH, to ask Robert B. Livingston, the Associate Chief of the Division of Research Facilities and Resources, to investigate the "Moral and Ethical Aspects of Clinical Investigation." But since the overwhelming sentiment in the Senate debate was to leave research unfettered, the Livingston Report, delivered in November 1964, was under little external pressure to recommend a change in the laissez-faire policy. The Report recognized that "There is no generally accepted professional code relating to the conduct of clinical research"; and expressed "a mounting concern ... over the possible repercussions of untoward events ... [because] highly consequential risks are being taken by individuals and institutions as well as NIH." But it did not urge the adoption of new policies. "There was strong resistance," recalled one of the participants, "on attempting to set forth any guidelines or restraints or policies in this area." The Report concluded that "Whatever the NIH might do by way of designing a code or stipulating standards for acceptable clinical research would be likely to inhibit, delay, or distort the carrying out of clinical research," and was, therefore, unacceptable (Frankel 1972: 20-22; Livingston Memorandum 1964).

But the Livingston Report was not the last word for incidents continued to surface. The most notorious case, the one that received the greatest publicity and whose impact was felt most directly at the NIH, was Chester Southam's cancer research with senile and elderly patients. The protocols, begun in 1963, captured considerable public attention over a two-year period – including a lawsuit, extensive press coverage, and a hearing before the New York State Board of Regents – almost all of which was hostile. "It made all of us aware," one NIH official later observed,

> of the inadequacy of our guidelines and procedures and it clearly brought to the fore the basic issue that in the setting in which the patient is involved in an experimental effort, the judgment of the investigator is not sufficient as a basis for reaching a conclusion concerning the ethical and moral set of questions in that relationship (Frankel 1972: 23, 31).

Just as the effects of the Southam protocols were being considered in the NIH, a well-publicized lecture that Beecher delivered in 1965 and then the appearance of his *NEJM* article demonstrated unequivocally that insensitivity to the ethics of experimentation was not unique to one investigator. At least one Congressman asked NIH directly how it intended to respond to Beecher's charges, and he was told that the findings "as might be expected have aroused considerable interest, alarm, and apprehension." Although "there are instances in the article which are either cited out of context, incomplete, or with certain mitigating circumstances omitted," still "constructive steps have already been taken to prevent such occurrences in research supported by the Public Health Service."[2]

The Congressman's letter was only the most visible sign of NIH's vulnerability (or sensitivity) to political and legal pressure. Washington officials who hoped to survive in office understood the need to have a policy on hand so that when criticism mounted, they could say that the problem *had* existed, but procedures were now in place to resolve it. Thus, NIH director James Shannon readily conceded that one of his responsibilities, even if only a minor one, was "keeping the Government out of trouble," and his advisers concurred: it would be nothing less than suicidal to believe that "what a scientist does within his own institution is of no concern to the PHS." Following immediately on these events, Shannon appointed an ad hoc group at NIH to consider policy changes, and it quickly concluded that should cases involving researchers' disregard of subjects' welfare come to court, the NIH "would look pretty bad by not having any system or any procedure whereby we could be even aware of whether there was a problem of this kind being created by the use of our funds" (Frankel 1972: 25).

But more than bureaucratic self-protection was at stake. The NIH response represented not only an exercise in self-protection but a reckoning with the substantive issues involved, with the causes behind the behavior of the individual researchers. By the mid-1960s, it had become apparent to the NIH leadership, and to many others as well, that conflict of interest marked the interaction of investigator and subject, that what was in the best interests of the one was not in the best interests of the other. The bedrock principle of medical ethics, that the physician acted only to promote the well-being of the patient, did not hold in the laboratory. The problem was structural and its roots lay in the fact, now recognized, that the doctor-patient relationship could not serve as the model for the investigator-subject relationship.

It was this perspective that moved Shannon and others at the NIH to alter its policies. Clinical research

> departs from the conventional patient-physician relationship, where the patient's good has been substituted for by the need to develop new knowledge, that the physician is no longer in the same relationship that he is in the conventional medical setting and indeed may not be in a position to develop a purely or a wholly objective assessment of the moral nature or the ethical nature of the act which he proposes to perform (Frankel 1972: 30).

The researcher's aims, in other words, were likely to pervert his ethical judgments. The intrinsic nature of his quest rendered him morally untrustworthy.

To compound matters, one could never be certain whether the investigator was seeking to promote the good of mankind or his own career. Although Beecher was most comfortable in blaming researchers' "thoughtlessness and carelessness" for protocols of dubious ethicality, he also cited the precipitous rise in NIH grant expenditures which both reflected and stimulated an all-consuming research ethos. "Medical schools and university hospitals are increasingly dominated by investigators.... Every young man knows that he will never be promoted to a tenure post ... unless he has proved himself as an investigator." Since "medical science has shown how valuable human experimentation can be in solving problems of disease and its treatment," the drive to produce led researchers to disregard the rights of their subjects. In fact, the drive to produce was so overdetermined, drawing on self-interest and scientific interest, that the investigator could no longer be trusted to act in anyone's best interest, save his own (Beecher 1966).

   The outcome of all these considerations was that in February 1966 and then in revised form in July 1966, the NIH established, through its parent body, the United States Public Health Service, guidelines covering all federally funded research involving human experimentation. The regulators moved very carefully, aware that they were in an unexplored and controversial territory. "This policy," explained Dr. William Stewart, the Surgeon General, in testimony before a Senate subcommittee, "seeks to avoid the danger of direct Federal intervention, case by case, on the one hand, and the dangers inherent in decision by an individual scientist on the other." The order of July 1, 1966 decentralized the regulatory apparatus, assigned "responsibility to the institution receiving the grant for obtaining and keeping documentary evidence of informed patient consent." It also mandated "review of the judgment of the investigator by a committee of institutional associates not directly associated with the project" and defined (albeit quite broadly) the standards that were to guide the committee: "This review must address itself to the rights and welfare of the individual, the methods used to obtain informed consent, and the risks and potential, benefits of the investigation." As Stewart explained:

> What we wanted was an assurance from [the grant-receiving institutions] that they had a mechanism set up that reviewed the potential benefit and risk of the investigation to be undertaken, and that reviewed the method that was used to obtain informed consent. And we thought that this should be done by somebody besides the investigator himself – a group. We thought this group might consist of a variety of people, and left it up to the institutions to decide.

Stewart was proud that

> We have resisted the temptation toward rigidity; for example, we have not prescribed the composition of the review groups nor tried to develop detailed procedures applicable to all situations.... Certainly this is not a perfect instrument. But ... this action has introduced an important element of public policy review in the biomedical research process.[3]

As important as it is to recognize the novelty of the NIH response, one must also appreciate the limits of policy changes designed and implemented by insiders. For one, the NIH did not mandate the inclusion in the collective decision-making process of those who were not part of the research world. In the 1966 policies, scientists reviewed scientists; peers, not outsiders, determined whether human subjects were adequately protected. Given its stance on conflict of interest, NIH did require that members of the review committee should have "no vested interest in the specific project involved." But it was ever so vague about other criteria, asking only that members should have "not only the scientific competence to comprehend the scientific content ... but also other competencies pertinent to the judgments that need to be made."

Second, and even more important, the NIH response focused far more on the review process than the consent process. It did recognize the importance of the principle of consent, changing the title of its Clinical Center manual from "Group Consideration of Clinical Research Procedures" (1953) to "Group Consideration and Informed Consent in Clinical Research" (1967). But NIH retained an investigator's skepticism about the ultimate value of the consent procedure, a position that was widely shared in the research community. For example, one of Beecher's Harvard colleagues, having read an early draft of the 1966 article, wrote him:

> Should informed consent be required? No! For the simple reason that it is not possible. Should any consent be required? Yes! Any teaching and research hospital must clearly identify itself as such ... to the patient upon admission.... The fact that the patient is requesting admission to this hospital represents tacit consent. How do we interpret tacit consent? Not as a license but as a *trust*. This adds not subtracts responsibility.[4]

In keeping with this sentiment, the internal memorandum that NIH enclosed with its new Clinical Center manual observed: "While there is general agreement that informed consent must be obtained, there is also the reservation that it is not possible to convey all the information to the subject or patient upon which he can make an intelligent decision. There is a strong feeling that the protection of the subject is best achieved by group consideration and peer judgment." Moreover, NIH was not yet prepared or able to be very specific about what such phrases meant in daily practice. "Many of these key terms," conceded Eugene Confrey, the Director of Research Grants, "lack rigorous definition or are incompletely defined for purposes of general application." And the problem was not semantic but conceptual; NIH officials did not fully grasp the implications of the consent principle. Thus, in explaining to "normal volunteers" at the Clinical Center that participation in the research was truly voluntary, they observed: "You will be asked to sign a statement in which you indicate that you understand the project and agree to participate in it. If you find your assigned project to be intolerable, you may withdraw from it."[5] Invoking the criterion of intolerability was hardly the way to educate subjects to the freedom of their choices.

Part of NIH's difficulties rested in the relative novelty of the principles of informed consent. Its precise meaning and implications were still obscure not only to research institutes but to courts. But part of the difficulties reflected the persistence of the notion that doctors should protect patients, not that patients should protect themselves. The NIH procedures of 1966 were innovative, but only to a point. Its leaders still reflexively expected professionals to ensure the well-being of the patient – and if they recognized the weaknesses of trusting to the individual researcher, they preferred to empower groups of researchers. The goal was to prevent harm to the subject, not to provide the subject with every opportunity and incentive to express his own wishes. The difference is substantial, and to move from one orientation to the other would require a different type of input from a different type of group, in essence, a shift from insiders to outsiders, a process that was in fact already underway.

III

As critical as the NIH initiatives were in setting a new precedent in the governance of physician-researcher behavior, the impact of the exposés in human experimentation cannot be measured or understood from its activities alone. One must leave Washington and the research establishment and look to broader societal considerations in order to answer such questions as: Why did the descriptions of the research protocols stir such public outrage? Why was it that the NIH came under external pressures to alter its procedures? It is altogether imaginable that the public response to the Southam protocols might have been lackadaisical, or better yet, to trust to the judgment of the researcher. If Doctor Southam said there was little or no risk, then there was little risk, and besides, the subjects of the experiments had no quality of life to speak of and might well have preferred to take this last opportunity to do a service to mankind. If Dr. Krugman declared that the retarded at Willowbrook would have inevitably contracted hepatitis and received better care in his nursing unit than on the wards, believe him and take comfort in the fact that since the subjects of the experiments had no quality of life to speak of, they might well have preferred to take this last opportunity to do a service to mankind. And on and on the rationales might go, with the majority happy to accept the sacrifice made by a minority so marginal to society. So why all the fuss, the anger, and the pressure to reform the system?

The answer begins with the changed political culture of the mid-1960s, particularly with the rights orientation that so dramatically reduced the discretionary authority of a number of social actors. Research scientists were only one of many groups to feel the effects of "a pervasive distrust of all constituted authorities." "The list of those who have suffered this loss," as I had occasion to note in a 1978 essay,

> is as lengthy as it is revealing: college presidents and deans, high school principals and teachers, husbands and parents, psychiatrists ... and obviously, prison wardens, social

workers, hospital superintendents, and mental hospital superintendents (Rothman 1978: 84-85).

There were substantial reasons for curbing the researchers' discretion – and the exposés provided them. Behind the more general distrust of constituted authorities lay a disillusionment with the idea that they would live up to the claim to exercise discretion in the best interest of others. The original purpose for the grant of power had been to allow them the room to carry out their benevolent designs, to substitute their more informed, enlightened and scientific judgments for those of their clients, children, students, and spouses. The research protocols that Beecher described and the newspaper stories repeated contradicted the assumption. Clearly, these investigators had not acted in the best interests of their subjects by injecting them with cancer cells, feeding them hepatitis viruses, or inserting cardiac catheters. Rather, the incidents confirmed the supposition of conflict of interest (the very conclusion that Beecher, Shannon and other insiders had reached). In a culture learning to think that deans would act in the best interest not of the student but of the university, that husbands would further their own needs, not those of their wives or children, that wardens thought first of the prison and psychiatrists of the mental hospital, and neither gave much weight to the inmate, it was not surprising to learn that research scientists also pursued their own priorities with little thought to their subjects' well-being. Self interest, career advancement and promotion – these motives were no less powerful in the laboratory than anywhere else, and no less properly contained and circumscribed. Claims to be acting for the betterment of mankind were too pat. It was a zero sum game both outside and inside the laboratory.

By the same token, the rights oriented culture of the 1960s fostered an extraordinary identification with the underdog, with the minority, as evidenced by the fact that it was the tactics of America's most important minority, the blacks, that became the model for others to emulate. Just as black activists used a language of rights to counter discrimination, so did advocates for women, children, gays, and students. It may not have been apparent or even correct that all of these groups constituted minorities or in any conventional meaning of the term possessed rights (in what sense does a child have a right against a parent?) but such qualifications did not slow the momentum of the movements. This orientation also affected the field of human experimentation, and gave a special meaning to the exposés. Was it not the case that the research subjects were very often drawn from minorities – the incarcerated, the poor, the handicapped, and the senile? Moreover, the incidents made the language of rights seem necessary and appropriate. If such protocols characterized clinical research, it was critical to define, and then to protect, the rights of research subjects – and, down the road, the rights of patients.

Finally, the eagerness to reduce discretion and secure rights led directly in human experimentation, as it did in the other areas, to a readiness to frame rules and regulations, to bring formality to procedures that heretofore had been casual and open-ended. In social welfare, the preference was to define entitlement rather than have the welfare mother trust to the discretion of the social worker; in delinquency and corrections, the goal was to promote procedural protections so that defendants and inmates did not have to rely upon the benevolence of a juvenile court judge or prison warden. And in this same spirit, in clinical research the aim was to establish a more exacting review process and a more formal and elaborate consent process so that the subjects did not have to rely for protection only upon the conscience of the individual researcher. All the movements presumed a conflict ("them" against "us") in which self-serving motives were cloaked in the language of benevolence, and majorities took every occasion to exploit minorities. In such a combative and competitive world, the best, really only protection, was to depend upon rules, not sentiment, to secure fairness.

It was not only lawyers preoccupied with rights and process but philosophers as well who transformed attitudes and policies toward human experimentation. Indeed, it was often the issues raised by human experimentation that helped bring philosophers into medicine. Paul Ramsey was certainly one case in point. *The Patient as Person* (1970) was undoubtedly the most significant work in the field after Fletcher's initial foray; and it is no coincidence that human experimentation – and some of the lessons that had been learned from it – was the opening focus of his analysis. The consent of children to medical investigations was the subject of the first chapter and Ramsey's working postulates came right out of the 1960s experiences. He was deeply concerned that the medical and scientific community acted on "the basis of an ethos all its own," and that ethos did not, he contended, provide sufficient protection for subjects (or patients). "I do not believe that either the codes of medical ethics or the physicians who have undertaken to comment on them ... will suffice to withstand the omnivorous appetite of scientific research or of a therapeutic technology that has a momentum and a life of its own." Indeed, he emphasized the critical nature of consent precisely because of "man's propensity to overreach his joint adventurer even in a good cause," a conclusion taken directly from human experimentation. Ramsey's interests did not stop with experimentation – his book goes on to discuss brain death, organ transplantation, and termination of treatment – but the key feature for us here is that they started with it.

Let one more example suffice for demonstrating how human experimentation served as the magnet bringing outsiders to medicine. The Spring 1969 issue of *Daedalus* was devoted to "Ethical Aspects of Experimentation with Human Subjects," the first time this broadly interdisciplinary publication had devoted an entire issue to medicine. Of the 15 contributors, 6 were drawn from the health sciences (including Henry Beecher), 5 from law (including Guido Calabresi and Paul Freund), and then one each from anthropology (Margaret Mead), sociology (Talcott Parsons), philosophy (Hans Jonas), and law and psychiatry (Jay Katz). Of course, some of these authors had already demonstrated a keen interest in bringing their disciplinary insights to medicine (Katz, Parsons), but some here were first entering a field where they would soon be doing important work

(Freund, Calabresi, Jonas). Crossing the bridge from one's home discipline to medicine had a somewhat scary quality to it. It is doubtful if any contemporary writer would be as circumspect as Jonas was in describing his initiation.

When I was first asked to comment "philosophically" on the subject of human experimentation, I had all the hesitation natural to a layman in the face of matters on which experts of the highest competence have had their say (219).

Since that time, ethics in medicine has become everyone's business. Laymen no longer hesitate to intervene. To an unprecedented degree, physicians now must share authority in medical decision-making. The controversy over human experimentation was a critical first step in this transformation.

NOTES

A fuller version of this argument and the sources for it will be found in my forthcoming (1990) book, *Ruling Medicine* (Pantheon).

1. Henry Beecher to Richard Field, August 3, 1965, Beecher Manuscripts, Countway Library, Harvard University.
2. John Sherman (associate director for the NIH extramural program) to Roman Pucinkski, July 1, 1966, NIH files, Washington, D.C.
3. Testimony of William Stewart before the U.S. Senate Subcommittee on Government Research, Committee on Government Operations, "Providing for the Establishment of the National Commission on Health Science and Society," Wednesday March 27, 1968, 211-212.
4. Henrik H. Bendixen to Henry Beecher, March 12, 1965, Beecher papers, Countway Library, Harvard University, Boston, Massachusetts.
5. NIH, "Handbook on the Normal Volunteer Patient Program of the Clinical Center (n.d., March 1967 stamp): 15.

REFERENCES

American Medical Association
    n.d.   Economics and the Ethics of Medicine. Chicago: Bureau of Medical Economics.
Beecher, Henry E.
    1966   Ethics and Clinical Research. New England Journal of Medicine 274: 1354-1360.
Blanchard, Paul
    1949   American Freedom and Catholic Power. Boston: Beacon Press.
Cabot, Richard C.
    1926   Adventure in the Borderlands of Ethics. New York: Harper and Brothers.
Daedalus
    1969   Special Issue: Ethical Aspects of Experimentation with Human Subjects.
Faden, R.R. and Beauchamp, T.L.
    1986   A History and Theory of Informed Consent. New York: Oxford University Press.
Fletcher, Joseph F.
    1954   Morals and Medicine; The Moral Problems of: The Patient's Right to Know the Truth, Contraception, Artificial Insemination, Sterilization, Euthanasia. Princeton: Princeton University Press.

Frankel, Mark S.

    1972   The Public Health Service Guidelines Governing Research Involving Human Subjects. Monograph 10, Program in Policy Studies in Science and Technology, The George Washington University. Washington, D.C.

Livingston, Robert B.

    1964   Memorandum to Director, NIH, Progress Report on Survey of Moral and Ethical Aspects of Clinical Investigation.

Ramsey, Paul

    1970   The Patient as Person: Explorations in Medical Ethics. New Haven: Yale University Press.

Rothman, David J.

    1978   The State as Parent. In W. Gaylin et al. (eds.) Doing Good: The Limits of Benevolence. New York: Pantheon Books.

    1987   Ethics and Human Experimentation: Henry Beecher Revisited. New England Journal of Medicine 317: 1195-1199.

Rush, Benjamin

    1947   The Vices and Virtues of Physicians (1801). In Dagobert D. Runes (ed.) Selected Writings of Benjamin Rush. New York: Philosophical Library. Pp. 293-307.

RENÉE C. FOX

THE EVOLUTION OF AMERICAN BIOETHICS:
A SOCIOLOGICAL PERSPECTIVE

"Bioethics" is a social and cultural, as well as an intellectual happening. The term came into use toward the end of the 1960s, in connection with an area of inquiry and action that was just beginning to develop in the United States at that time. What has since become the flourishing field of bioethics is structured around problems associated with modern biomedicine, particularly some of its frontier scientific and technological advances. But bioethics is not just the name of a new discipline. It also refers to a wider, progressively more organized, professional and public concern about these so-called ethical matters that has taken on some of the contours and characteristics of a social movement.

Bioethics, in both these senses, surfaced in American society in the late 1960s, in a period of acute and ramifying social and cultural ferment. From its outset, the value and belief questions with which it has been preoccupied have run parallel to those with which the society has been grappling more broadly. Its name, empirical focus, and technical vocabulary notwithstanding, the matters it treats are not exclusively medical and ethical. They have more general, moral and social, and also religious connotations.

> Bioethics is not just bioethics,... and [it] is more than medical [we have written].... Using biology and medicine as a metaphorical language and a symbolic medium, bioethics deals ... with nothing less than beliefs, values, and norms that are basic to our society, its cultural tradition, and its collective conscience (Fox and Swazey 1984: 336, 360).

As the field has evolved over a period of what will soon be two decades, it has attracted greater public notice and participation, more media attention, and increasing government interest and involvement. Although its overall configuration and core foci have remained relatively constant, bioethics has undergone several shifts that are congruent with those taking place on the larger American scene.

BIOMEDICAL FOCI

Bioethics has concentrated its attention on a particular group of advances in biology and medicine. With strikingly little acknowledgment of the improvement in identifying, controlling, and treating disease that these advances represent, bioethics has focused on actual and impending problems that medical scientific progress has brought in its wake:

developments in genetic engineering and counseling, life support systems, birth

201

G. Weisz (ed.), Social Science Perspectives on Medical Ethics, 201–217.
© 1990 *Kluwer Academic Publishers. Printed in the Netherlands.*

technology, population control, the implantation of human, animal, and artificial organs ... [and] the modification and control of human thought and behavior are principal foci of [bioethical] concern. Within this framework, special attention is concentrated on the implications of amniocentesis ... *in vitro* fertilization, the prospect of cloning,... organ transplantation, the use of the artificial kidney machine, the development of an artificial heart, the modalities of the intensive care unit, the practice of psychosurgery, and the introduction of psychotropic drugs (Fox 1974: 446-447).

Cross-cutting its interest in these areas of biomedical development is the strong involvement in issues of human experimentation that the field of bioethics has shown from its inception. In fact, it was growing professional and public concern about the moral conditions under which experimentation with human subjects was being conducted in our society, particularly in the sphere of medical research, that played the major triggering role in the genesis of bioethics. During the field's early years (from the late 1960s to the mid-1970s), the greatest interest and energy were invested in this area. Bioethical analysis and action were centered on the general importance and difficulty of obtaining the informed voluntary consent of human research subjects (and also of patients who were not subjects) for the procedures they underwent, and on the worrisome situations of potential subjects who were especially vulnerable, disadvantaged, or unable to speak for themselves (such as infants and children, those who are mentally ill, retarded, or incompetent, the dying, persons like prisoners and the mentally infirm, who are institutionalized, uneducated and very poor individuals, and members of minority groups that have been the object of much societal prejudice and discrimination).

Human experimentation continued to be a central preoccupation of bioethics, and remains so to this day. However, in the mid-1970s, concern about life and death and personhood issues at the beginning and end of the life cycle, began to take up more medical, philosophical, and legal space in bioethical discussion. In this, its second phase, bioethics became increasingly involved in the proper definition of life and death and personhood, and with the humane treatment of emerging life, and life that is passing away, particularly with the justifiability of foregoing life-sustaining forms of medical treatment.

The mid-1980s constitute a third phase in the evolution of bioethics: its progressive "economization." This current "ethics and the new economics of health care" juncture not only focuses on conceptual and practical aspects of the economic framework within which health care is being delivered in the United States, but also on the same allocation of scarce resources and cost containment themes that the current political administration stresses.

Along with the several thematic shifts through which American bioethics has passed, there have been fluctuations in the prominence accorded to particular medical scientific advances and types of medical treatment. For example, in the mid-1970s, when discussion of the case of Karen Ann Quinlan was at its height,[1] and again, in the 1980s, at the peak of concern about so-called Baby Doe cases,[2] adult and neonatal intensive care units, and various life support modalities received a great deal of bioethical notice. Or, to take another example, a resurgence of interest in cardiac transplantation has occurred in the mid-1980s, in synchrony with the increase in transplants that the immunosuppressive drug cyclosporine has helped to effect, the temporary, bridge-to-transplantation clinical trials of the artificial heart that have been attempted, and the creation of a National Task Force on Organ Transplantation in 1985, by an act of the U.S. Congress. Such shifts and fluctuations in the salience of different bioethical themes and topics result from a complex process of interaction between concrete medical scientific and clinical events, the settings and the circumstances in which they occur, professional and public reactions to them (including those of the media and the polity), and the work agendas of the interdisciplinary community of "bioethicists." The subject matter with which bioethicists are involved at any given time is partly self-determined, and partly responsive to current biomedical happenings, and to the phenomena and questions that they are invited or commissioned to analyze by a wide range of groups – professional organizations, medical institutions, academic milieux, private foundations, government agencies and commissions, legislative bodies, the courts, and the media, among them.

SOCIAL ORGANIZATION AND INSTITUTIONALIZATION OF AMERICAN BIOETHICS

As the foregoing suggests, American bioethics now functions within a complex, multi-institutional framework, that encompasses both private and public, and local and national sectors of the society. Its initial organization toward the end of the 1960s, however, was much simpler, and largely confined to the activities of the pioneering centers and programs that established the field. Preeminent among these were the Institute of Society, Ethics, and the Life Sciences in Hastings-on-Hudson, New York (The Hastings Center), the Center for Bioethics, of the Kennedy Institute of Ethics, Georgetown University, Washington, D.C., and the Society of Health and Human Values, initially located in Philadelphia, and now based in McLean, Virginia. These three groups have continued to play an important leadership role in the intellectual development of bioethics and its involvement in social action. But in the course of the last fifteen years, they have been joined by scores of other independent, academic, professional, and public interest associations, institutes, departments, and programs that have a major commitment to reflection, research, teaching, publishing, and/or action in matters pertaining to bioethics. These burgeoning organizations have received sponsorship and support from a wide array of private foundations, scholarly bodies, and government agencies.

The expansion and increasing institutionalization of bioethics have also contributed to "a veritable explosion of literature on bioethical issues" (Walters 1983, 9: 3). This literature is so vast, and appears in such diverse and widely scattered sources that a number of bibliographies have been published to facilitate its use. The best known of these is the *Bibliography of Bioethics*, edited by LeRoy

Walters, Director of the Georgetown University/Kennedy Center for Bioethics, that has been published annually since 1974, and that seeks to be as comprehensive as possible for all English-language bioethical materials.[3] There is also an *Encyclopedia of Bioethics*, published in 1978, under the editorship of Warren T. Reich who, like LeRoy Walters, is a scholar associated with the Georgetown/Kennedy Center for Bioethics.[4] As Reich himself commented in his Preface, "[I]t is unusual, perhaps unprecedented, for a special encyclopedia to be produced almost simultaneously with the emergence of its field."

One of the most remarkable features of the way in which bioethics has developed is the extent to which it has pervaded the public domain. American courts and legislatures, at every level of the system, are involved with bioethical questions – deliberating on them, rendering decisions about them, and formulating guidelines, regulations, and laws that bear upon them. Issues concerning research with human subjects and the availability and foregoing of life-saving and life-sustaining medical treatment have been particularly important nuclei of this legislative and judicial action.

There have also been two consecutive national commissions that have been convoked to study and report on a number of bioethical problems in medicine and research. The first of these was the National Commission for the Protection of Human Subjects of Biomedical and Behavioral Research, created by the National Research Act passed by Congress in July 1974, and appointed by the Secretary of Health, Education and Welfare. This Commission was charged with the task of conducting "a comprehensive examination and study to identify the basic ethical principles which should underlie the conduct of biomedical and behavioral research involving human subjects," and of developing "guidelines which should be followed in such research to assure that it is conducted in accordance with such principles." From 1975 through 1978, the Commission published a series of reports on various aspects of experimentation with special categories of human subjects from whom it is problematic to obtain informed consent: human fetuses and pregnant women, children, prisoners, mentally ill or retarded persons who are institutionalized, and individuals who are possible candidates for psychosurgery. The recommendations presented in these reports provided the bases for the regulations governing human experimentation adopted by the Department of Health, Education, and Welfare, and by its successor, the Department of Health and Human Services. The second commission, the President's Commission for the Study of Ethical Problems in Medicine and Biomedical and Behavioral Research, was authorized by Congress in November 1978, and its original members were named by President Jimmy Carter. (I was one of these members.) Its mandate was broader than that of the National Commission; it was not confined to ethical issues in research with human beings, and it was accorded the flexibility to alter, or add to the list of topics it was asked to study, either as it, or the President saw fit. The Commission began its work in January 1980, and was officially terminated at the end of March 1983. It published reports on making health care decisions, securing access to health care, defining death, deciding to forego life-sustaining treatment, screening and counseling for genetic conditions, genetic engineering, compensation for injured research subjects, whistleblowing in biomedical research, and it also issued two biennial reports required by its Congressional mandate, on protecting human subjects, and implementing human research regulations. In addition, it collaborated with the National Institutes of Health's Office for Protection from Research Risks, and with the Food and Drug Administration to develop a guidebook for Institutional Review Boards. A number of these reports have had considerable influence on public opinion and public policy, medical and hospital practice, and on legislative action, and legal decision-making. The Commission's volumes on *Defining Death* (July 1981), and on *Deciding to Forego Life-Sustaining Treatment* (March 1983) have probably had the most impact.

The "going public" of bioethics has been amplified by the continuous, extensive, and prominent coverage that the issues with which it deals have received from the print media and television. Media interest in bioethically relevant events and questions has escalated over the years, spiking locally and nationally to especially high points when there are identified human cases that dramatically personify the medical questions, and the societal values and beliefs that are involved.

There is, however, a certain inflation in the public and professional notice being given to bioethical questions. A kind of "everything is ethics" syndrome has developed. It consists of a generalized tendency to attach the label "bioethical" to a diverse array of medical science and technology-associated problems that are moral, social, and religious in nature, as well as strictly ethical. Accompanying this trend is the propensity to call all intellectuals and professionals lecturing and writing about value issues relevant to medicine, "ethicists" – whether they are philosophers, religionists, biologists, physicians, lawyers, or social scientists. In turn, these patterns are part of the "bioethics bandwagon" process that now exists on the American scene.

Its expansionary characteristics notwithstanding, certain kinds of concerns and ways of thinking about them germane to the area of bioethics, are defined as outside its orbit, or as peripheral to the field. This is all the more striking because of how engulfing the scope of bioethics appears to be. What is included and excluded in American bioethics, and emphasized and deemphasized by it, can be better understood through an analysis of its ethos.

THE ETHOS OF AMERICAN BIOETHICS[5]

The chief intellectual and professional participants in American bioethics, and shapers of its ethos have been philosophers (particularly philosophical and religious ethicists), theologians (predominantly Christian), jurists, physicians, and biologists. Certain public officials who have played a prominent role in the involvement of local and national government in bioethical matters have also influenced the outlook and emphases of the field. In addition, the thought and

presence of economists have been more strongly felt in bioethics during the
1980s, as allocation of scarce resources and cost containment problems have
become more salient. In comparison, the participation of anthropologists, political scientists, and sociologists has been quite limited. This is a complex
phenomenon, resulting as much from the prevailing *Weltanschauung* of present-
day American social science as from the framework of bioethics. (I will return to
a further consideration of the relationship between bioethics, social thought, and
sociology.)

The values and beliefs highlighted by American bioethics represent a particular cross-section of the society's cultural tradition. From the outset, the conceptual framework of bioethics has accorded paramount status to the value-complex
of individualism, underscoring the principles of individual rights, autonomy,
self-determination, and their legal expression in the jurisprudential notion of
privacy. By and large, what bioethics terms "paternalism" is negatively defined,
because however well-meaning, and concerned with the good and the welfare of
another person it may be, it interferes with and limits an individual's freedom and
liberty of action.

The notion of contract has played a major role in the way that relations
between autonomous individuals are conceived in bioethics. Self-conscious,
rational, functionally specific agreements between independent individuals are
presented as ethical models. The archetype of such contractual relations is the
kind of informed, voluntary consent agreement between subjects and investigators in medical research that the field of bioethics helped to formulate and
translate into public policy.

Veracity and truth-telling is another value-precept that is stressed by bioethics.
In keeping with the overall orientation of the field, bioethics attaches special
importance to the rights of patients or research subjects to "know the truth" about
the discomforts, hazards, uncertainties, and "bad news" that may be associated
with medical diagnosis, prognosis, treatment and experimentation.

As already indicated, concern about the just and fair distribution of scarce,
expensive resources for advanced medical care, research, and development is
another major value-preoccupation of bioethics, and one that has increasingly
come to the forefront in the 1980s. The view of distributive justice underlying it
is structured around an individual, rights-oriented conception of the general or
common good, in which greater importance is assigned to equity than to equality.
"Cost containment," an essential value-component of this outlook on rightful
distribution, has evolved from the status of an ethical response to an empirical
situation of economic scarcity, to something approaching a categorical moral
imperative of bioethics.

Finally, what is usually referred to as "the principle of beneficence" or
"benevolence" is also a key value of bioethics. The enjoinder to "do good" and
"avoid harm" that this entails is structured and limited by the supremacy of
individualism. The benefiting of others advocated in bioethical thought is
circumscribed by respectful deference to individual rights, interests, and auton-

omy; and minimizing the harm done to individuals is more greatly accentuated than the maximization of either personal or collective good.

The weight that bioethics has placed on individualism has relegated more socially-oriented values and ethical questions to a secondary status. The concept and the language of *rights* prevails over those of *responsibility, obligation*, and *duty* in bioethical discourse. The skein of relationships of which the individual is a part, the socio-moral importance of the interdependence of persons, and of reciprocity, solidarity, and community between them, have been overshadowed by the insistence on the autonomy of self as the highest moral good. Social and cultural factors have been primarily seen as external constraints that limit individuals. They are rarely viewed as forces that exist *inside*, as well as outside of individuals, shaping their personhood and enriching their humanity.

The restricted definition of "persons as individuals," and of "persons in relations" that dominates bioethics has downplayed "values like decency, kindness, empathy, caring, devotion, service, generosity, altruism, sacrifice, and love" (Fox and Swazey 1984: 355). All of these involve "emphasis on ... feeling, on connection and relatedness" (Keller 1985: 173),[6] and on responding to known and unknown others in a self-transcending way.

The basic values of bioethics, and its cognitive characteristics and style of thought are closely allied. The capacity to think rationally and logically about ethical questions, with as much rigor, clarity, consistency, parsimony, and objectivity as possible, is accorded high intellectual and moral standing. Bioethics proceeds in a largely deductive manner, formalistically applying its mode of reasoning to the phenomenological reality it addresses. An array of cognitive techniques are used to distance and abstract bioethical analysis from the human settings in which the questions under consideration occur, to reduce their complexity and ambiguity, and to control the strong feelings that many of the medical situations on which bioethics centers can evoke in those who contemplate them, as well as those who live them out.

Much of the bioethical literature is based on the assumption that the value questions which have arisen in the field of biomedicine have been "caused" or "created" by medical scientific and technological advances. Partly because of its biomedical and technological determinism, bioethical analysis does not usually take note of the fact that some of the same cultural questions that have crystallized around biological and medical developments have also been central to many nonmedical issues that have surfaced in American society during the past fifteen to twenty years. Thus, the examination of bioethical matters is enclosed within a framework that does not easily open onto a consideration of larger, more encompassing concerns about the general state of ideas, values, and beliefs in American society at this time. In fact, there is a sense in which bioethics has taken its American societal and cultural attributes for granted, ignoring them in ways that imply that its conception of ethics, its value system, and its mode of reasoning transcend social and cultural particularities. In its inattention to its "American-ness," and its assumption that its thought and moral view are *trans*cultural, American bioethics has been more intellectually provincial and chauvinistic than it has recognized.

This inclination, along with the dominance that bioethics assigns to individual ethical questions, bend it away from involvement in social problems. Nowhere is this more apparent than in the context of the neonatal intensive care unit (NICU), where bioethical attention has been riveted on the justifiability of nontreatment decisions, but relatively little attention has been paid to the fact that a disproportionately high number of extremely premature infants of very low birth weight, with severe congenital abnormalities, cared for in NICUS are babies born to poor, disadvantaged mothers, many of whom are single teenagers, and also nonwhite. Bioethics has been disinclined to regard the deprived conditions out of which such infants and mothers come as falling within its purview. These are defined as *social*, rather than ethical problems – a dichotomous distinction that permeates bioethical thinking. In effect (to use a term coined by a philosopher-founder of bioethics, Daniel Callahan), these kinds of social problems are "de-listed" as ethical problems in a manner that removes them from the sphere of moral scrutiny and concern.

Bioethics deals with religious variables in a comparable fashion. When questions of a religious nature arise in bioethics, there is a tendency either to screen them out, or to "reduce" them, and fit them into the field's circumscribed definition of ethics and ethical. This is particularly notable with regard to issues concerning what sort of life-sustaining medical treatment, if any, should be used to preserve the lives of severely mentally or physically defective, or terminally ill persons. The "strictly ethical" manner in which bioethics generally defines and approaches such an ultimate question contrasts with the way in which it has been understood and dealt with in American courts. Judges have repeatedly acknowledged that these medical and moral problems are also "spiritual," "theological," and "metaphysical," as they have put it. For this reason, mindful of the American doctrine of separation of church and state, they have tried to identify the religious facets of the questions before them, and have consistently advised that any government intervention into these "difficult and sensitive" decisions "should ... reflect caution and sensitivity."[7] ("Courts temporal are not ideally suited to resolve problems that originate in the spiritual realm"(Kaufman 1985).[8])

These cognitive and value traits of bioethics are a product of a number of converging factors. American philosophy and philosophers have had the greatest molding influence on the field. It is principally American analytic philosophy – with its emphasis on theory, methodology, and technique, and its utilitarian, neo-Kantian, and "contractarian" outlooks – in which the majority of the philosophers most active in bioethics were trained. Their philosophical positivism is reinforced by the principles and rules of "being scientific" that physicians and biologists have been educated and socialized to apply to their own professional work, and that they have brought to bioethics. In turn, the rationalism of American law, its emphasis on individual rights, and the ways in which it has been shaped by Western-American traditions of natural law, positivism, and utilitarianism overlap with, and enhance key attributes of the philosophical and scientific thought in bioethics. As has been indicated, the law has shown less of a tendency than analytic philosophy to collapse medical, moral, social, and religious issues into an undifferentiated amalgam called ethics. Nevertheless, in order for the courts, and also the legislatures to make concrete decisions about the swelling numbers of bioethical cases and problems that have been coming before them, rather than declaring them non-justiciable, or beyond the application of legal principles, they, too, must logically "reduce" the questions involved, and "technicalize" them. Thus, for example, issues concerning the "definition of death" are transformed into statutory, medico-legal criteria for pronouncing death, either on the basis of irreversible cessation of circulatory and respiratory functions, or irreversible cessation of all functions of the entire brain;[9] rather than "speculate" on the answer to "the difficult question of when life begins," raised by abortion, the courts turn to the biological concept of "fetal life after viability," and to the legal concept of "right of personal privacy";[10] and "failure to follow procedural requirements in [their] promulgation," becomes the primary basis on which Baby Doe rules are declared invalid by the Court, its broader commentary on life-sustaining medical treatment for newborn infants who are gravely ill or have serious congenital defects, notwithstanding.[11]

In the philosophical, medical, and legal spheres of bioethics, applied pragmatism also plays an important role in how bioethical problems are conceptualized and analyzed. Physicians, nurses, other medical professionals, hospital administrators, patients, families, biologists, lawyers, judges, legislators, politicians, business executives, and their associates are called upon to make up their minds about what to do and what not to do in real-life settings, and then act on the basis of their determinations. In addition, intellectuals and academicians who are considered to be bioethics experts are often asked to help professional practitioners and policymakers arrive at reasonably specific and clear ways of resolving concrete problems that they face. The practical necessity of systematically managing the complexity and uncertainty that such decision-making and applied action entail augments the cognitive predisposition of bioethics toward utilitarianism, positivism, and reductionism.

Despite the significant contributions of highly esteemed religious ethicists and theologians to bioethics, the field is studiously secular in its perspective. This secularism is partly a consequence of the professional socialization that philosophers, biologists, physicians, and jurists undergo in their respective fields. It is also an instrumental, political, and moral response to a basic societal question that the whole phenomenon of American bioethics poses: How can, and should, an advanced modern, highly individualistic, pluralistic, and religiously resonant society, like the United States, founded on the precept of governance "under law," rather than "under men," and the sacredly secular principles of separation of church and state and freedom of belief, try to achieve collective and binding consensus about the kinds of bioethical issues that are now in the public domain? The society is experiencing great procedural as well as substantive difficulty in resolving, on behalf of its entire citizenry, these more-than-medical ethical matters that lie at the heart of its moral, religious, and cultural tradition. The problem is complicated, and made more acute by the degree to which such questions have entered the polity. Siphoning off their religious content, and framing them in as secular way as possible provides an institutionally supported, reductionistic way of defining them, compatible with the ethos of bioethics, that makes them more amenable to logical analysis and technical solution. The problem is that, in the end, this masks their essential nature, and because this is true, does not conclusively dispel them.

American bioethical thought and discourse have two other distinctive characteristics. The field reflects the fact that its major intellectual shapers and spokespersons have been professionals, scholars, and academics. The literature and commentary it has generated tend to be "locked into ... upper middle class professional and guild enclaves." The relatively few bioethicists who have taken the initiative of going to local communities to discuss ethical and humanistic questions associated with health, illness, and medicine, have found that the "grassroots" public "knows the issues," but knows them "in their own language."[12]

Finally, in a number of respects, the ideological orientation of bioethics could be characterized as conservative. The way that bioethics has defined individualism, the centrality that it has accorded to this value-complex, the degree to which it has played down a social perspective on personal and communal moral life, its parsimonious insistence on a cost-containing framework of analysis, and the extent to which its rationality and methodology have distanced it from the phenomenological reality, and the human complexity of lived-in medical ethical situations, have converged to form a gestalt that is congruent with other fundamentals of a conservative outlook.

INCIPIENT CHANGES IN THE ETHOS OF AMERICAN BIOETHICS

A certain amount of intellectual and moral stock-taking has been occurring in the American bioethical community since 1980, as a consequence of the experience and greater maturity of insight it feels it has gained from its early years of development, and its interaction with medical, legal, media, and public policy milieux. Self-criticisms, as well as critical commentaries by observers of bioethics, have started to appear in print.

One of the major foci of the auto-critique in which bioethics has begun to engage is what is referred to as its "excessive emphasis" on the "language of 'rights'" of American individualism and of American courts (Callahan 1980: 1230), and its elevation of the principle of "autonomy" to a position of such overriding importance that it was "given a kind of moral clout sufficient to trump every other value" (Callahan 1984: 42). Most philosopher-bioethicists still contend that "autonomy's temporary triumph" (Veatch 1984: 38-40) (and even its triumphalism), were "indispensable" for general moral, and specifically medical reasons:

> Among ... the benefits of giving moral priority to autonomy ... are a recognition of the rights of individuals and of their personal dignity; the erection of a powerful bulwark against moral and political despotism; a becoming humility about the sources or certainty of moral claims and demands; and a foundation for the protection of unpopular people and causes against majoritarian domination ... (Callahan 1984: 40).

> The issues and themes of the medical ethics that emerged in the 70s were largely shaped by the problems bequeathed to us by clinicians and patients of the day ... that led us to see medical ethics as a conflict between the old Hippocratic paternalism (having the physician do what he or she thought was best for the patient) and a principle of autonomy ... (Veatch 1984: 38).

The issue that is being raised by bioethicists in the 1980s is "not whether and how to get rid of autonomy." Rather, it is how to keep it from becoming such "a moral obsession" that it "pushes other values aside," particularly those that pertain to *social* ethical questions, "a search for morality in the company of others, community as an ideal, and interdependence as a perceived reality" (Callahan 1984: 42).

Admonitions about an "ethic based on maximizing individual autonomy at the expense of personal obligation to the human community – past, present and future" (Morison 1984: 43, 48), constituted the major motif at the June 1984 symposium on "Autonomy, Paternalism, and Community," that the Hastings Center organized to celebrate its fifteenth anniversary. Repeated affirmations were voiced about the necessity to progress beyond a "minimalist," individualistic ethic based on the proposition that "one may morally act in any way one chooses so far as one does not do harm to others" (Callahan 1981: 19-22). Central to all the symposium discussions was a preoccupation with "the just allocation of scarce resources" that was cited as "perhaps the dominant theme of bioethics today" (Gaylin 1984: 5), and with the fact that because of its strong individualistic focus, ethics had contributed relatively little to such large-scale, societal, health care delivery and policy issues. The statements made on this important rites-of-passage occasion seemed to presage greater future involvement of bioethics in what were termed "large, structural, moral and political decisions."

In connection with the future agenda of bioethics, several other items are increasingly mentioned. One of these is the felt necessity to make a greater effort to overcome "the current problems with medical ethics thinking and its application in medicine," that philosopher-psychiatrist Colleen D. Clements calls "bioethical essentialism." The first step in this direction that she advocates is "the acceptance of the uniqueness and diversity of cases, which implies situation-relative definitions of action guides. It is most unethical to routinely apply policies and procedures equally to cases," she states (Clements 1985: 204).

Of special interest to sociologists are recent statements that bioethicists have made about the importance of learning to work more closely with those in the social sciences, as well as in medicine and the law, as the field becomes more intensively involved with social, cultural, and cross-cultural dimensions of moral life, the actual experience of patients, families, and caretakers of different backgrounds in the daily round of medicine, and with policy issues that are national and international in scope. The Kennedy Institute of Ethics, for example, has established an Asian Bioethics Program (directed by a Japanese Doctor of Law), that describes itself as "a resource and reference center in the United States on the issues relating to ethics and values caused by the impact of scientific, technological, and biomedical development in Asian countries," and that serves as a "think tank" for bioethical issues to the Institute of Medical Humanities, School of Medicine, Kitasato University where, in August 1985, and again in August 1986, it cosponsored the First and Second "Japan-U.S. Symposium for Bioethics." The *Bibliography of Bioethics*, published by the Kennedy Center for Bioethics, is making plans to include gradually French-Canadian, French, Dutch, German, and Japanese materials in its citations. And, in collaboration with the Georgetown Center for Population Research, and the Hebrew Home of Greater Washington, the Kennedy Center has embarked on an action-research project concerned with "frail elderly" persons in long-term care, that includes four social workers on its staff.

RELATIONS BETWEEN AMERICAN BIOETHICS AND SOCIOLOGY

Such discernible shifts in a more social and cultural direction notwithstanding, the incorporation of social science and scientists into bioethics has not proceeded very far. The August 1986 issue of the *Hastings Center Report* is illustrative. It features four articles on "Caring for Newborns," written by an Israeli, an Indian, a Japanese, and an American author respectively. These articles emphasize and describe how the social structure, cultural values, religious traditions, and world-views of the eight different societies that they cover, their respective stages of socioeconomic development, and their polities affect their "strategies" of caring for "imperiled or impaired" newborn infants. Another major article in the same issue (Purtilo and Sorrell 1986) summarizes and discusses a 1984-1985 study that was made of the ethical dilemmas that confront physicians practicing in an American rural setting, and how they are like and unlike those of their urban counterparts. The article highlights the influence that distance from other professional facilities, the interrelationship of private and professional roles in a small community, and the non-specialized orientation of their practice has on what rural physicians consider to be ethically justifiable courses of professional action under sociomedical circumstances that "sometimes force them to 'march to the beat of a different drummer' than their city colleagues do." Despite the social and cultural subject matter and perspective of these articles, none of them is written by, or cites a social scientist. Two of the authors are physicians; one is a medical student; two are jurists; and one is a scholar in the field of medical jurisprudence and humanities.

The relations between bioethics and social science continue to be tentative, distant, and also susceptible to strain. On the bioethics side, the ethos of the field has still not evolved far enough from its individualism and autonomy-of-self starting point to accord the kind of recognition to the principles of relatedness, reciprocal obligations, and community, or to what sociologist Emile Durkheim termed the "non-contractual," fiduciary framework of social life, to be firmly convinced that social scientific competence can be useful in the analysis of medical ethical phenomena. (Here, it is relevant to note that, although the Kennedy Institute of Ethics' project on the "frail elderly" entails collaboration with demographers, and *in situ* interventionist research involving nursing home patients, family members, and social workers, its chief goal is to "facilitate the expression of autonomy" by older persons in "long-term care decision-making.")

One of the inadvertent consequences of the continuing professional dominance of philosophy and philosophers in the field is the tendency on the part of influential members in the "invisible college" of bioethicists to regard the social sciences as disciplines whose subject matter and value concerns are alien to those of the humanities, and to view most social scientists as insufficiently humanistic in their education and their perceptions. The overall trend is to equate social science with the "amassing" of quantitative data and information, relevant to "policy formulation and implementation," using large-scale surveys and "the techniques of linear programming and of cost-benefit, risk-benefit, and systems analysis" (Callahan 1980: 1233) to do so. At the same time, social science is seen as less rigorously scientific than biomedicine. Social scientists who work in a more ethnographic, *en plein air*,[13] qualitative fashion are often defined by philosophically-oriented bioethicists as "closet humanists," who are "not really" social scientists. Although usually, this is considered to be a compliment by those who make the allegation, it often carries with it the implication that such social scientific humanism-in-disguise is more prone to "relativism" and "subjectivism" and to "moral laxity and logical confusion" (Geertz 1982: 22) than the discipline of philosophical ethics is.

This tension that exists in bioethics goes beyond the "science vs. humanities," and "two cultures" divide that C. P. Snow delineated (Snow 1959). Something like a "three cultures split" is involved, with "the 'Third Culture' Snow forgot" being the social sciences that fall in a rather negatively evaluated "intermediary sort of ... sub-area" (Geertz 1982: 27). Numerous physicians, biologists, and jurists, as well as moral and religious philosophers active in bioethics share this perspective, which is institutionalized across the primary disciplines that have fashioned the matrix of bioethics. In this connection, it is significant that the great expansion in the teaching of bioethics to medical students and house staff that has occurred in American medical schools and medical centers in recent years, has progressively displaced the behavioral science teaching about psychological, social, and cultural dimensions of health, illness, and medicine that was prominent in the 1950s and 1960s. In most instances, the fact that medical ethics has been substituted for social science in the curriculum has neither been deliberately planned by medical educators, nor recognized by them as such.

The limited involvement of social scientists in bioethics, however, is not solely attributable to the unwelcoming ambivalence with which they are viewed by the shapers and gatekeepers of the field. The ethos of the social sciences also contributes to the minor role that they have played in bioethics. Most sociologists, for example, have not chosen to concentrate on the kinds of problems with which bioethics is concerned; and many are unaware either that bioethics exists, or of its potential sociocultural import. This is related to sociologists' greater propensity to work in a social structural or social organizational frame of analysis than to focus on systems of values and beliefs – the very cultural variables that are core to bioethics – and to their positivistic preference for quantitative rather than qualitative methods of research. In turn, these intellectual inclinations and disinclinations are reinforced by an ideological conviction that is held by a substantial number of sociologists, especially those who espouse a critical, reformist, or radical approach to the field – namely, that dealing with culture and cultural tradition, which are slow and difficult to change, is inherently more conservative than being interested in what are assumed to be more malleable and rapidly modifiable social structures. Along with these factors that have curtailed sociologists' participation in bioethical phenomena and issues, the lack of pertinent interdisciplinary competence in the sociological profession has been another deterrent. Some ability to handle relationships between social and cultural variables, on the one hand, and biomedical, philosophical, and/or legal considerations, on the other, is requisite. There is a dearth of sociologists who have this trained competence, or who seem willing to acquire it.

The branch of sociology that one would expect to have the most affinity for an analysis of social and ethical implications of biomedical advance and technology is the sociology of medicine. And indeed, among the relatively few sociologist authors who have contributed to the bioethics literature, most are known for their work on medicine. But in this subspecialty, too, sociologists have been less interested in areas that bear directly on the concerns of bioethics, such as the sociology of medical science, medical research, therapeutic innovation, and the nexus between the sociology of medicine and of law, and more involved in such matters as the "sick role," and "illness behavior" of patients, the "professional dominance" and "organized autonomy" of physicians, the process of medical socialization, and the social system of the hospital. The field's stronger commitment to "basic" academic than to "applied" work also acts as a deterrent to sociologists' participation in bioethics.

Still, the fact remains that the kinds of studies and analyses that sociologists have made of illness, medical care, the hospital, and the medical professions are germane in many respects to the situations and value questions that are central to bioethics.[14] It is the ethos of the two fields, and the way that they converge to produce reciprocal blindspots, that seems to prevent both sociology and bioethics from recognizing this. One would like to think that this collection of essays goes some way towards overcoming such mutual disciplinary isolation.

NOTES

This essay appears in somewhat different form in Fox 1989.

1. In 1976, the New Jersey Supreme Court granted Joseph Quinlan's request to be appointed guardian of his daughter, Karen Ann Quinlan, who had been in a "persistent vegetative state" for more than a year, and empowered him to discontinue the life-supporting respirator on which she was being maintained, if attending physicians and the hospital ethics committee agreed that there was "no reasonable possibility" that she would return to a "cognitive, sapient state." This court decision overruled the previous position taken by Karen Ann Quinlan's physicians, and overturned the Superior Court of New Jersey's ruling on the case in 1975. Karen Ann Quinlan was taken off the respirator, but lived for almost ten more years, breathing on her own, without returning to consciousness. [See *In re Quinlan* 70 N.J. 10 355 A.2d 647, *cert. denied* 429 U.S. 922 (1976).]

2. The term "Baby Doe cases" has grown up around instances in which, out of a process of consultation between medical professionals in a neonatal intensive care unit (NICU) and parents, the decision has been made to withhold some medical or surgical treatment vital to maintaining the life of an infant. The Baby Doe-type cases that have received the most bioethical and public attention have involved infants afflicted either with Down's syndrome or spina bifida, who have both correctable life-threatening defects, and also serious, irreparable, permanent handicaps. Around such Baby Does, highly publicized court cases have occurred, precipitated in part by the direct intervention of the federal government which, at the specific request of the White House in 1962, led the Department of Health and Human Services (HHS) to establish so-called "Baby Doe rules," and at one point, a "Baby Doe hotline," to govern the NICU situation, and to report persons suspected of withholding medical or surgical treatment or nutrition from an infant in the NICU to the HHS.

3. The *Bibliography of Bioethics* is one of the ongoing research projects of the Kennedy Institute of Ethics. It has received financial support for its publication from the Joseph P. Kennedy Jr. Foundation, the National Library of Medicine, and the National Institutes of Health.

4   The *Encyclopedia of Bioethics* project was made possible by a grant from the National Endowment for the Humanities. The Endowment matched gifts from non-federal sources, including the Joseph P. Kennedy Jr. Foundation, the Raskob Foundation, The Commonwealth Fund, The Loyola Foundation Inc., and The David J. Greene Foundation Inc.

5. Parts of the discussion of the ethos of bioethics in this paper are heavily dependent on Fox and Swazey 1984: 352-358.

6. This phrase is taken from Evelyn Fox Keller's depiction (1985: 173) of Nobel Laureate in Medicine, Barbara McClintock's "vision of science." It is interesting, and more than coincidental that these concepts and values which, as Keller points out, coincide with "our most familiar stereotypes of women," are deemphasized both in the dominant ethos of science and of bioethics.

7. In the United States District Court for the District of Columbia, American Academy of Pediatrics, National Association of Children's Hospitals and Related Institutions, Children's Hospital National Medical Center, Plaintiffs, *v.* Margaret M. Heckler, Secretary, Department of Health and Human Services, Defendant, Civil Action No. 83-0774, filed 14 April 1983, U.S. District Judge Gerhard H. Gesell presiding.

8. Irving R. Kaufman is a judge on the U.S. Court of Appeals for the Second Circuit.

9. See the Uniform Determination of Death Act, drafted by the President's Commission for the Study of Ethical Problems in Medicine and Biomedical and Behavioral Research (1981: 119-120) in collaboration with the American Bar Association, the American Medical Association, and the National Conference of Commissioners of Uniform State Laws, and the review and analysis of

state legislation on the "determination of death," and judicial developments in the "definition" of death in the same volume (109-146).

10. Supreme Court of the United States, Roe et al. *v.* Wade, District Attorney of Dallas County, Appeal from the United States District Court for the Northern District of Texas, No. 70-18. Argued December 13, 1971; reargued October 11, 1972; decided January 22, 1973.

11. American Academy of Pediatrics, National Association of Children's Hospitals and Related Institutions, Children's Hospital National Medical Center *v.* Margaret M. Heckler, cited above in note 7.

12. Personal letter (July 17, 1986), from Ruel Tyson, Professor in the Department of Religious Studies of the University of North Carolina at Chapel Hill.

13. This is a phrase used by anthropologist Clifford Geertz (1982: 18) in his Bicentennial Address to the American Academy of Arts and Sciences.

14. Relevant sociological work of this sort is discussed in Fox 1989: 229-266.

RERERENCES

Callahan, Daniel
  1980 Contemporary Biomedical Ethics (Shattuck Lecture). The New England Journal of Medicine 302: 1228-1233.
  1981 Minimalist Ethics. Hastings Center Report 11, 5: 19-25.
  1984 Autonomy: A Moral Good, Not a Moral Obsession. Hastings Center Report 14, 5: 40-42.
Clements, Colleen D.
  1985 Bioethical Essentialism and Scientific Population Thinking. Perspectives in Biology and Medicine 28: 188-207.
Fox, Renée C.
  1974 Ethical and Existential Developments in Contemporaneous American Medicine: Their Implications for Culture and Society. Milbank Memorial Fund Quarterly, Health and Society 52: 445-483.
  1989 The Sociology of Medicine: A Participant Observer's View. Englewood Cliffs, N.J.: Prentice Hall.
Fox, Renée C. and Judith P. Swazey
  1984 Medical Morality is Not Bioethics – Medical Ethics in China and the United States. Perspectives in Biology and Medicine 27: 337-360.
Gaylin, Willard
  1984 Introduction: Autonomy, Paternalism, and Community. Hastings Center Report 14, 5: 5.
Geertz, Clifford
  1982 The Way We Think Now: Toward an Ethnography of Modern Thought. Bulletin of The Academy of Arts and Sciences 35: 14-34.
Kaufman, Irving R.
  1985 Life and Death Decisions. The New York Times, October 6.
Keller, Evelyn Fox
  1985 Reflections on Gender and Society. New Haven, Conn.: Yale University Press.
Morison, Robert S.
  1984 The Biological Limits on Autonomy. Hastings Center Report 14, 5: 43-49.
President's Commission for the Study of Ethical Problems in Medicine and Biomedical and Behavioral Research
  1981 Defining Death. Medical, Legal and Ethical Issues in the Determination of Death. Washington, D.C.: U.S. Government Printing Office.
Purtilo, Ruth and James Sorrell
  1986 The Ethical Dilemmas of a Rural Physician. Hastings Center Report 16, 4: 24-28.
Reich, Warren T. (ed.)
  1978 Encyclopedia of Bioethics. Vols. 1-4. New York: The Free Press/Macmillan.

Snow, C.P.
   1959   The Two Cultures and the Scientific Revolution (The Rede Lecture, 1959). New York: Cambridge University Press.
Veatch, Robert M.
   1984   Autonomy's Temporary Triumph. Hastings Center Report 14, 5: 38-40.
Walters, LeRoy (ed.)
   1974-1988 Bibliography of Bioethics. Vols. 1-6. Detroit: Gale Research Co. Vols. 7-9. New York: The Free Press/Macmillan. Vols. 10-14. Washington, D.C.: Kennedy Institute of Ethics.

PART IV

MEDICAL ETHICS AND SOCIAL SCIENCE

RICHARD W. LIEBAN

MEDICAL ANTHROPOLOGY AND THE COMPARATIVE STUDY OF MEDICAL ETHICS

Medical anthropology has emerged relatively recently as a recognized field of study. Its development in the past few decades has been rapid and comprehensive. If we exclude significant work in the field by physical anthropologists – in such areas as paleopathology, forensic medicine, health aspects of population genetics, the scope of research in medical anthropology by cultural anthropologists is formidable. Among the topics that could be cited are the studies of a wide variety of medical systems which examine nosologies, etiologies, modes of diagnosis and therapy, characteristics of medical personnel, practitioner-patient relationships, and assessments of efficacy; medical pluralism; illness and health-seeking behaviors; social and cultural aspects of epidemiology; anthropological studies of nutrition, demography and obstetrics; cultural psychiatry; alcoholism and drug abuse; applied anthropology in public health programs and clinical settings; and economic and political dimensions of health problems and health care.

Largely lacking as a focus of interest in medical anthropology has been medical ethics. For example, if we examine each of seven general review articles, published between 1953 and 1983, that have summarized research developments, directions and findings in the field, references to medical ethics are absent or negligible (Caudill 1953; Polgar 1962; Scotch 1963; Fabrega 1972; Lieban 1973; Colson and Selby 1974; and Landy 1983).

The relatively little attention that has been given to medical ethics by medical anthropologists is particularly accentuated against a background of generally heightened interest in medical ethics in recent years; this is reflected in such phenomena as a substantial growth of literature in the field (Brody 1976), a striking increase in the number of courses in medical schools devoted to medical ethics (Veatch 1978), and the establishment of various institutes and centers. A number of factors may help explain why there has not been more concern with medical ethics in anthropology.

1. Value judgments are a basic feature of much of the literature on medical ethics. But the influence of cultural relativism has had a constraining effect on value judgments relating to the ethics of non-Western medical systems, the study of which has been a major focus of medical anthropology. There are substantial differences of view among anthropologists regarding cultural relativism (Spiro 1986). However, as it applies to field research in societies other than that of the researcher, cultural relativism has in some form been characteristic of anthropology (Pelto and Pelto 1973; Hatch 1983). This has meant studying other cultures on their own terms and avoiding ethnocentric value judgments about them. Value judgments aside, there has been a tendency, consistent with relativistic methodology perhaps, to view the ethical aspects of health care in other cultures as cultural

G. Weisz (ed.), Social Science Perspectives on Medical Ethics, 221–239.
© 1990 Kluwer Academic Publishers. Printed in the Netherlands.

givens and to neglect ways in which they may relate to moral questions and ambiguities.

2. The non-Western medical systems on which, at least until recently, anthropologists have concentrated much of their attention are not those in which recent revolutionary developments in biomedical knowledge and technology have occurred. These developments, by creating new medical options, and in combination with socio-economic changes, have engendered new kinds of moral issues and dilemmas, or added new dimensions to old ones. They have contributed significantly to the contemporary growth of interest in bioethics and medical ethics in Western nations (Kass 1985). Most non-Western medical systems studied by medical anthropologists have not been subject to comparable changes with their impact on ethics (Bulger 1987).

Regardless of the reasons why medical ethics has received relatively little attention in medical anthropology, there are indications that the situation is changing. Kunstadter (1980) has called for comparative anthropological research on medical ethics in cross-cultural and multi-cultural perspective. Although there has been relatively little published so far in this vein, such recent cross-cultural studies as those dealing with child abuse and neglect (e.g., Korbin 1981; Scheper-Hughes 1987) involve significant moral issues. The increased attention of medical anthropology to Western biomedicine (Chrisman and Maretzki 1982; Gaines and Hahn 1982; Hahn and Gaines 1985; Lock and Gordon 1988) has also in some cases been marked by concern with ethical questions. Relevant in this respect is a textbook in medical anthopology by Foster and Anderson (1978), who devote a chapter to bioethical aspects of birth, old age and death in American society. The authors examine critically American assumptions and practices regarding key phases of the life cycle and contrast them with comparable characteristics of certain non-Western societies. Recent field studies by medical anthropologists that include consideration of certain ethical aspects of Western biomedicine will be discussed later in this paper.

Some thirty years ago a sociologist friend of mine jokingly said that the difference between sociology and anthropology was that sociologists studied societies with dry cleaning. Of course, that was a hyperbole then and it would be more so today when it is even less true that anthropologists can be sorted out from other social scientists on the basis of the level of development of societies that each studies. Still it is true that anthropologists have devoted proportionately more of their research to non-Western societies than have other social scientists and that a comparative approach to human society and culture has been a characteristic of anthropology.

One of the contributions that medical anthropologists can make to the study of medical ethics at this point is to provide more extensive knowledge about the similarities and differences in moral aspects of medical standards and behavior in different cultures. Prominent in the field of medical ethics is discussion of general ethical principles relating to health and medicine, including the grounds for validating them, and their applicability to actual situations and issues (Jonsen and

Hellegers 1974; Beauchamp and Childress 1983; Veatch 1981; Kass 1985; Bulger 1987). In contemporary medical ethics, these principles have largely been discussed in relation to Western biomedicine and its social environment. But the study of ethical principles in medicine should be accountable to a wider representation of human thought and experience than that found in Western biomedicine.

In this respect, we need to know far more than we do about what might be called the "ethnoethics" of medicine in non-Western societies. This would include moral norms and issues in health care as understood and responded to by members of these societies. Ethnoethics should be informative not only about cross-cultural variation in ethical principles of medicine, but also about variations in the issues which in different societies become defined as morally relevant or problematic. Ethnoethical information should contribute to the discourse on medical ethics not only by illuminating culturally distinctive moral views and problems, but also by helping to provide a more realistic and knowledgeable basis for the exploration of cross-cultural ethical similarities.

In this paper, much of the emphasis will be on ethnographic description and analysis of morally relevant aspects of medical actions and issues in particular cultural settings. A major purpose is to exemplify some of the types of problems that might be considered in anthropological studies of medical ethics.

MEDICAL ETHICS OF FOLK HEALERS

Let me begin with data culled from my field notes which were reexamined for their ethical content. These focus on folk healers, called *mananambal*, in Cebuano areas of the Philippines. Cebuanos are one of the major Christian lowland groups in the Philippines, and *mananambal* usually gain legitimacy for their healing role and derive their healing power from their spiritual connections with or sponsorship by Biblical figures.

In Cebuano folk medicine, the morality of the *mananambal* is considered to have a vital bearing on his own well-being and on his effectiveness as a healer. The practitioner's character before he gets the spiritual call to assume the healer's role is apparently irrelevant to his selection. He may have a reputation as a "good" person, be morally undistinguished, or be a reprobate. The selection process itself seems to be a case of Divine mystery. But the call itself carries with it moral expectations of the recipient that may confront the *mananambal*-designate with a severe personal crisis.

Healers themselves believe that the *mananambal* may become insane or physically sick if he does not meet moral obligations that go with the call. One described his fear and how he wept when St. Joseph, his special sponsor, first appeared and told him he should treat the sick. He said that going through his mind at the time were strong doubts about his ability to give up such vices as running around with women, drinking and gambling, vices that could subject him to insanity after he became a *mananambal*. A married man, he was particularly afraid that he would not be able to resist the temptation of extra-marital affairs to which he had grown accustomed.

Meeting moral obligations is not only important for safeguarding the *mananambal*'s health, but also for maintaining his healing power. As the *mananambal* discussed above put it, "If a *mananambal* sins, he will go down, get weak, lose courage, so he can no longer penetrate the hardness of life." The point is further illustrated by a quarrel between the wife of another *mananambal* and his mistress, in which the wife said, "You are the cause of my husband's patients not coming to him anymore. He has lost the power of healing because of you."

Fundamental among his obligations are those of the healer to his patients. Characteristically, *mananambal* are told by their spiritual sponsors that they have the duty to help the sick from all walks of life. In addition, *mananambal* and others refer to the healer's moral obligations in spheres of life that go well beyond the clinical setting and relations with patients. Whether or not the *mananambal*'s adherence to certain moral obligations influences his behavior with patients, it protects his healing power and ultimately serves his patients' needs. The healing power is a gift to be employed for the benefit of the sick.

Viewed cross-culturally, the inititation of the *mananambal* by a call from the spiritual world that is often resisted, as well as punishment from a spiritual source if the healer violates obligations, are characteristic of shamans and other practitioners whose healing power derives from spirits (e.g., Fabrega and Silver 1973; Kleinman 1980; Foster and Anderson 1978). Such features of the relationship between the healer and his spiritual sponsor signify control of the relationship by the sponsor.

Some of the most important problems in the ethics of Western biomedicine involve communication between practitioners and patients including informed consent, veracity and confidentiality (Katz 1984; Beauchamp and Childress 1983). I would like to devote the rest of this section to ethical issues related to communication between *mananambal* and their patients, beginning with confidentiality.

Kunstadter (1980) has pointed out the contrasting ethical implications of a medical system in which therapy requires public disclosure and community action from one in which the practitioner has the obligation to protect the patient's privacy. Even when curing rituals that require some kind of community participation are not employed as part of therapy, treatment in public is a common feature of folk medical systems (Foster and Anderson 1978; Kleinman 1980), and this is applicable to *mananambal* and their patients.

Discussions in public about their symptoms did not appear to disturb patients of *mananambal* whom I observed, but public disclosures about the patient's behavior could be another matter. For example, the *mananambal* described above, whose sponsor was St. Joseph, emphasized the need for rectitude by patients if they were to preserve their health or have it restored. Frequently, he reproached them in front of others, sometimes strangers, for such transgressions as living together outside marriage or failure to attend church. Sometimes, particularly when the patient was a friend, he did this in a bantering way, but at other times he could be a stern lecturer. Then his remarks were not sermons to anonymous or collective "sinners," but highly personal revelations about identified individuals who were in the presence of others. To illustrate, once, and this was not unusual, I heard him ask a grim-faced woman how she expected to be cured when she made sacrifices to spirits outside the pale of God, did not confess to a priest or go to church regularly, and practiced favoritism among her children.

Care must be exercised not to make unwarranted assumptions about the emotional impact on those reproached in such circumstances, but there were indications of strong feelings on the part of some patients. Once the healer admitted that, perhaps, at times he went too far in his criticism of patients. As an example, he told about a young unmarried woman who had come as a patient; he had said in front of other patients that she was not a virgin, a statement that could seriously shame her in a society which strongly emphasizes the importance of pre-marital chastity among females (Lieban 1967). She had never returned, and once had crossed the street to avoid him. Another *mananambal*, would, sometimes at a patient's request, sit at a somewhat remote table with the patient and whisper about personal matters that the patient did not want publicly divulged. So, there were indications that some patients regarded the experience or prospect of public disclosure of personal matters as something to be avoided. If this were the case, it apparently did not deter many from becoming or remaining patients of the first *mananambal* who had one of the largest followings of any *mananambal* in Cebu City at the time.

In certain ways, the actions of this *mananambal* could be regarded as ethically justified, supportive of moral values of his society, particularly those emphasized by the church, and necessary for modifying the behavior of patients so they could be cured. However, in some respects the ethics of his disclosures could be viewed as problematic, considering how painful the experience might be for some patients, although this could be outweighed by their quest for health.

This situation is quite different from that, say, described by Turner (1964), in which an important therapeutic ritual among the Ndembu of south central Africa reflects beliefs in the relationship between social conflict and illness: members of the community voluntarily participate in a curing ceremony where those engaged openly confess antipathies that they and the patient feel toward one another; the emphasis is on reconciliation as a means of restoring health. But the Cebuano cases I have discussed suggest that the ethics of the situation may be complex and ambiguous when public disclosures about patients, a common feature of folk healing, are involved.

I want to turn now to another aspect of communication between *mananambal* and patient. Veracity is a multi-faceted principle of the ethics of practitioner-patient relationships; these facets range from strictures against and problems concerning lying and deception to duties regarding disclosure. Usually disclosure is discussed primarily with reference to information the practitioner possesses about the patient that is relevant to the patient's right to know and that could affect his welfare. However, such information can have wider impact on others as well, and in what follows I am going to consider the effects of disclosure on both the patient and others.

Cebuano society is not sorcery-ridden. But sorcery is one of the causes of illness according to folk etiology. I collected data on more than 100 sorcery cases, in most of which the *mananambal* either identified the putative instigator of the attack or confirmed the suspicions of the patient.

If a healer makes a sorcery diagnosis and identifies the person responsible for the attack, his disclosure of this information to his patient can be ethically justified on the grounds that the healer has provided the patient with information the latter is entitled to know about his illness and its source. However, several *mananambal* said that in cases where they knew sorcery was responsible for the patient's illness, they would not tell the patient, or would at least claim that they did not know who had perpetrated the sorcery, because such information could lead to serious trouble, notably the seeking of vengeance by the patient. Suspicions or accusations of sorcery can deeply embitter social relations, and while I was doing research in the Philippines newspapers occasionally carried stories about the slaying of an individual suspected of sorcery or witchcraft (Lieban 1967). Under these circumstances, the ethics of disclosure has to be viewed in a context wider than the patient's right to medical information and the healer's obligation to provide it.

ETHICS IN THE POPULAR HEALTH CARE SECTOR

Kleinman (1980: 50) speaks of the popular sector as the largest but least studied part of any health care system. He defines the popular sector as the "lay, non-professional, non-specialist popular culture arena in which health is first defined and health care initiated." Although the study of medical ethics has tended to focus on the ethics of medical professionals, the major importance of the community's role in medical ethics has been noted in the literature (Kass 1985; Veatch 1981). In medical anthropology, the significance of lay beliefs, attitudes and actions in health care has been emphasized by Kleinman, who, for example, found in a study in Taiwan that 93 per cent of all illness episodes were first treated in the family, and 73 per cent were treated only in the family (Kleinman 1980: 182-83). Janzen (1978), in a study of the BaKongo of Zaire, has convincingly delineated the crucial importance of what he calls the "therapy managing group," composed of both kin and non-kin, in health care decisions and the delivery of health services. The topic of lay medical ethics is potentially one of great scope, but within the limits of this paper I will only suggest and exemplify a few of the types of problems that warrant further investigation.

Kunstadter (1980) sees problems of the allocation of scarce medical resources as an issue in medical ethics that probably recurs in all societies but which can be handled in different ways. It cannot be assumed, however, that, in every case, the determination of priorities for the distribution of limited resources that affect health will necessarily be viewed as an ethical issue from within the society.

For example, higher valuation of sons than daughters is widespread in South Asia, and there is substantial evidence that this can influence differences in health care and life chances of children (Beals 1976; Ramanamma and Bambawale 1980; Koenig and D'Souza 1986). Thus in a study of 228 villages in Bangladesh, Koenig and D'Souza (1986) found significantly higher mortality rates for female than male children, a phenomenon they related to better nutrition and health care for males. This reflects resource allocations that in their view result primarily from the interaction of economic and cultural factors. To illustrate how cultural factors combined with economic stringencies to influence the sex differential in mortality, they note that this differential first emerges systematically during the period when breastfeeding becomes inadequate to meet the nutritional requirements of a child, and parents are forced to make selective decisions about the allocation of such scarce and costly resources as supplementary food. As they see it, the strong cultural bias in favor of males is likely to be a central consideration in determining priorities for such allocations.

Such a pattern of allocating scarce resources basic to health and survival would raise serious ethical issues in the light of the principle of distributive justice that has an important place in the literature of Western medical ethics (Beauchamp and Childress 1983; Veatch 1981; Engelhardt 1986). Beauchamp and Childress (1983: 184) define distributive justice as "the justified distribution of benefits and burdens in society." This brief and broad definition, in itself, does not address significant questions and differences of view. But what is significant here is that Western medical ethics has come to define the manner of distributing resources that affect the maintenance or restoration of health as a moral problem.

However, at least on the basis of the information we have, we cannot be sure whether, to what extent, or how the distribution of food and medical care that favors male over female children involves moral dilemmas and choices for the Bangladesh villagers studied. Does distribution ever involve perceived conflict between the higher valuation of males than females and other values of the villagers? From the perspective of the villagers, is the distribution that favors male children a self-evident economic necessity that raises no moral issues? It is worth noting that in rural Bangladesh, residence is patrilocal, and property, particularly land, is owned by men (Cain 1977). A rigid division of labor obtains, so that the physical seclusion of women excludes them from most agricultural production and limits their economic contribution to the household. In contrast, the productive labor of sons is fundamental to the economy of the household, and sons may by early adolescence become net producers for the household (Koenig and D'Souza 1986: 15). These authors go on to say that "sons represent one of the few, and perhaps the most certain, forms of security – both in old age, and more immediately, against the risk of status-threatening events such as the death or incapacitation of the head of household." Under the circumstances, do Bangladesh villagers ever translate economic need into moral justification for the greater investment of health-related resources in sons than daughters?

Decisions discussed above about the allocation of health-related resources in rural Bangladesh were made within the family, generally the basic social component of the popular health care sector. In this case, family actions influencing members' health were consistent with traditional cultural values. But adherence to family role obligations, as these obligations are defined in terms of traditional values, can be increasingly difficult in societies undergoing rapid culture change. And this can be a source of ethical difficulties.

Conflict between family health care obligations and other role obligations or interests is epitomized in societies where adult children have traditionally had binding responsibility for the welfare of aging parents, including health care, but where the effects of modernization are now straining or eroding those commitments. Korean society is a case in point. As observed by Koo and Cowgill (1986), the aged in Korea, at present have very low priority when it comes to the national allocation of programs and resources. The government depends on the family both to provide personal care and to finance health care for the elderly. Consistent with this policy, government officials recently have been reminding younger family members of their obligations under the ethic of filial piety to care for elderly parents.

Indications are that younger Korean adults still strongly feel those obligations. One study of people living in three-generation households found that three out of five adult children acknowledged their responsibility to care for incapacitated parents; another study found over 90 per cent of the care of bedridden elderly patients was being provided by members of the family. However, current changes in the Korean family make such care increasingly difficult and burdensome. Young people now strongly prefer to live apart from their parents, and increasing numbers of them migrate from their communities of origin, especially from rural communities to cities, leaving elderly parents behind. Another trend that creates role dissonance is increasing employment of women outside the home. Such women can be torn between responsibilities as employees, mothers and dutiful daughters or daughters-in-law. Under the circumstances described by Koo and Cowgill for Korea, traditional ethical values of family health care can be transmuted into painful ethical dilemmas.

Thus far in this section, we have considered moral principles and issues essentially in the context of the family. However, the family may sooner or later seek help from practitioners, at which point, the ethics of health care involves the views and actions of the medical specialist and his lay clientele. Again taking the family as an illustration, the nature and scope of obligations assumed by the family with respect to the health care of its members can have an important bearing on the ethical perspective of the practitioner as well as the patient.

To illustrate, Kleinman (1980: 281) describes how both Western-style and Chinese-style doctors in Taiwan explain virtually nothing to the patient when they manage life-threatening illness. "They aim their remarks at the family members whose job it is to support the patient. For example, they tell the family, not the patient, that he is suffering from cancer or that the treatment of a serious illness has been unsuccessful. This is viewed neither by patients nor by practitioners as an ethical dilemma." While bypassing the patient when transmitting information about his illness is apparently not an ethical issue in these circumstances, it would be, inter alia, in relation to a major principle discussed in the literature of medical ethics, that of autonomy (Beauchamp and Childress 1983; Engelhardt 1986; Kass 1985; Veatch 1981), which in its most general, unqualified meaning refers to a condition where the individual is self-governing. Much of the discussion of autonomy in the Western literature would have little relevance to the medical context in Taiwan described by Kleinman, where the family, rather than the patient, commonly makes crucial decisions about the treatment of serious illness.

ETHICS OF BIOMEDICINE IN CROSS-CULTURAL PERSPECTIVE

A cross-cultural approach to the study of biomedical ethics does not imply that medical anthropologists should, following their traditional concerns with other cultures, concentrate more or less exclusively on non-Western settings and leave Western biomedicine to others. Biomedicine in the West is fundamental to the comparative study of medical ethics, and in recent years there have been indications of the sorts of contributions that anthropologists can make in this regard. A few examples are briefly discussed below.

Questions of ethics play a part in a study by Katz (1985) of decision-making among surgeons. She describes how surgical decisions are influenced by non-medical variables, such as collegiality, various organizational and bureaucratic factors, competition between specialties, increased referrals and notions of appropriate income. As Katz shows, medical decisions compromised by such non-medical considerations can have serious consequences for patients.

Cassell (1987), who has also studied American surgeons, finds that they value decisiveness, control and certitude. She argues that these qualities, which are selected for and reinforced in surgical training, are adaptive for surgery. In her view, this does not mean that every manifestation of arrogance, lack of consideration and insensitivity from the profession should be countenanced. However, she argues, surgeons should be criticized, at least partially, on their own terms, an approach that anthropologists have tended to adopt only when studying down (disadvantaged peoples and groups). In her view (Cassell 1987: 244):

> It is important, from a very practical point of view, that we show the same consideration when we study up. If we want to understand powerful groups, in order to meliorate or alter aspects of their behavior, we must understand who they are, why they are that way and (on occasion) how it is to our benefit that they are that way.

As noted previously, issues related to the use of new biomedical technologies have become major concerns of medical ethics. Koenig (1988) uses her study of a recent medical technology, therapeutic plasma exchange, to illustrate how a "technological imperative" to use a dramatic new device if it exists becomes "transformed into a moral imperative to provide a new therapy." Koenig describes how social factors support this process, and she notes that evaluations by physicians of the safety and efficacy of new techniques are highly subjective and not based solely on scientific criteria. Under the circumstances, patients make decisions about receiving treatment on the basis of technological assessments that may be biased or seriously flawed. As Koenig (1988: 490) sees it, "supported by the activist orientation of western physicians ... the centrality of research goals, and reimbursement policies biased toward the use of equipment-embodied technologies, the decision to use a new technology seems inescapable. In the process true patient autonomy is usurped."

One basis for interest by medical anthropologists in Western biomedicine is that it is not simply a biotechnical system which responds mechanically to and literally reflects realities of the natural world that affect health. Rather, Western biomedicine, like other forms of medicine, is a cultural system, a product of human society that encompasses knowledge interpretive of the natural world, as well as attitudes, values and patterns of communication and social relationships (cf. Hahn and Kleinman 1983; Gordon 1988; Lock 1988). This cultural system, including its ethics, can be studied at various levels and from different perspectives.

Research by anthropologists in clinical settings is conducive to ethnographic intensity, and ethical issues in such domains can be illuminated by such traditional fieldwork methods as participant observation, structured and unstructured interviews, case studies, and situational analyses. (For an incisive sociological study employing such methods and relevant to the study of ethics, see Bosk 1979.) Anthropologists not only study clinicians, they can assume clinical roles themselves. One who did so, Alexander (1979), suggests that anthropologists as clinicians must make moral judgments; but they confront the dilemma of whether these judgments should be based on the standards of those within the medical system or on other considerations, perhaps even moral absolutes, relevant to all systems.

Young (1982) stresses the need to examine social dimensions of medical phenomena and their extension beyond the clinic. He calls for a medical anthropology that includes research on how social forces and institutions of the larger society shape sickness and the healing process. Studies along these lines obviously involve description and analysis on a different scale than clinically focused ethnography. Ethically relevant macro-analysis has characterized recent work in critical medical anthropology that has among its basic concerns the impact of the political economy and power relations on biomedical institutions (e.g., Baer, Singer and Johnsen 1986).

Before considering the influence of cross-cultural differences on biomedical ethics, mention should be made of intracultural variations and their impact. A case in point would be ethnic variation in a society such as the United States, where such factors as differences between cultural assumptions of practitioners and patients and contrasts between class orientations can significantly impair delivery of ethnically appropriate health care services (Harwood 1981). Health deficits of ethnic minorities and frequent inadequacies in their health care raise serious moral issues that clearly fall within the purview of an anthropology of biomedical ethics.

Kunstadter (1980) offers as a working hypothesis that apparent agreement on medical ethics across cultures is a consequence of the diffusion of Western ethics together with Western medicine, a hypothesis supported by the fact that in India ethical codes of Western physicians and a range of other medical personnel, including Ayurvedic and Unani practitioners, are virtually identical and clearly modeled on the British paradigm. Kunstadter goes on to observe that in fact non-Western systems of medical ethics have been largely ignored by the Western-influenced medical establishments in non-Western countries. While this may be true, it does not mean that the ethics of biomedicine are unaffected by cross-cultural variations in values and social realities. As an illustration, we can compare differences in ways physicians in Israel and South Asia respond to imperiled or impaired newborn infants.

Eidelman (1986), a physician, reports that in Israel a number of factors militate against discontinuation of treatment for the afflicted infant. According to Eidelman, these include: religious beliefs about the sanctity of life (concepts such as euthanasia, even passive, are unacceptable to both religious and secular communities); the experience of the Holocaust and the role of German physicians in selecting "defective" individuals and groups have resulted in widespread belief in the sacredness of life, regardless of its defects and limitations; government funding for even the most complex intensive newborn care and for continuing care of the handicapped, freeing the individual family from these costs; and the concern of Jews that they will be outnumbered by the increasing Arab population. As Eidelman describes it, "Zero population growth is a nonexistent term in Israel and when one is concerned about numbers one is not selective; all children are welcome."

In contrast, Subramanian (1986), also an M.D., describes decisions about neonatal care in India, Sri Lanka, and Nepal based on criteria that differ substantially from those which seem to be utilized in Israel. According to Subramanian, in cases where a physician is attending the afflicted child, except when surgery is required, physicians make decisions completely autonomously. Their decisions are guided by what Subramanian calls "quality of life considerations." In addition to medical factors, these may include family composition, the family's ability to obtain further care, and cost factors. Also considered are limited resources which might be used to better effect on another patient with greater potential for a "good" quality of life. Fatalistic beliefs color physicians' decisions, and Subramanian states, "Quality of life rather than sanctity of life is a consideration because of a strong belief in rebirth."

In this comparison, the ethics of biomedicine appear in each case to be consistent with basic values and beliefs of the practitioners' society. But in other instances, serious ethical issues can arise that entail fundamental contradictions between biomedical perspectives and the established norms and values of the practitioner's society.

To illustrate, female circumcision is practiced in a number of African countries (Gruenbaum 1982; Gallo 1986). It may take the form of clitoridectomy, or a more drastic procedure, infibulation. Girls' health, and in some cases lives, are at risk at the time of circumcision, when hemorrhage, septicemia, retention of urine or shock may occur, while infibulation, by obliterating natural openings, can pose serious health hazards.

In Somalia, midwives circumcise, particularly in rural areas; but currently there is more demand for the operation to be performed by medical or paramedical personnel who, working in hospitals, can use drugs and more hygienic practices. Under the circumstances, the opinions which future medical personnel hold about female circumcision could eventually have an important bearing on the practice, and Gallo (1986) conducted surveys to obtain this information from samples of 58 female and 30 male medical students and 144 female nursing students (the latter due to replace unlicensed midwives in performing circumcision in rural areas). Gallo notes that despite efforts in school to inform students about the negative aspects of circumcision, 51 per cent of nursing students and 21 per cent of medical students found some positive aspects to the practice.

Gruenbaum (1982), in a paper on female circumcision in Sudan, argues that an effective policy to eliminate the practice must be broadly oriented toward changing the status of women and solving the social problems that affect them. One of the key points that Gruenbaum makes concerns the social cost to women if they were not to be circumcised. She says,

> Women in Sudan generally derive their social status and economic security from their roles as wives and mothers. Among most cultural groups in northern Sudan, female virginity at marriage is considered so important that even rumors questioning a girl's morality may be enough to besmirch the family honor and to bar her from the possibility of marriage. In this context, clitoridectomy and infibulation serve as a guarantee of morality.

Similar cultural assessments are made by male medical students in Gallo's Somalia survey. Eighty-one per cent of them considered the practice of female circumcision injurious. However, 50 per cent found some advantages to the process, primarily as it relates to religion rather than to sexual activity or hygiene; 38 per cent favored continuation of the practice, and 29 per cent said they would want their own daughters to be circumcised. Most of them said their own families would regard a son marrying an uncircumcised woman as tantamount to his marrying a prostitute and certainly as an act opposed by the Moslem religion.

In this case, health and medical considerations, at least as they have influenced some future medical personnel, are a basis for critical views toward traditional practices related to social and religious values, defiance of which could have dire personal and social consequences. And ambiguities in medical ethics could be related to (1) the extent to which future practitioners perceive conflict between acceptable standards of health and medicine and practices based on traditional social and religious values, and (2) the powerful grip of these practices on social behavior.

CROSS-CULTURAL SIMILARITIES

Comparative studies in the social sciences are concerned with description and analysis of social and cultural differences and similarities. To this point we have given considerable attention to variation in medical ethics and ethical issues. Now we would like to concentrate, at least briefly, on similarities. It should be noted that the discussion is concerned primarily with approximating similar phenomena across cultures, particular manifestations of which may vary considerably.

To focus the discussion, I am going to confine comparison to a few examples from two medical systems: Western biomedicine, whose ethics are the subject of a voluminous literature, and traditional medicine in Imperial China, whose ethics are examined in an important and informative work by Paul Unschuld (1979). Since the two medical systems are the products of different cultural traditions, we are dealing with separate cases rather than one case writ large. Therefore, resemblances found between them should be attributable to factors that independently foster comparable ethical features.

Unschuld relates the history of medical ethics in China to the struggle to gain control over medical resources and to the development of professionalization as a primary means to that end. In his view, stress on professional ethics in China was initiated by independent physicians, with such ethics "designed to persuade the public that whoever is in control and possession of medical resources uses them in a morally trustworthy manner" (Unschuld 1979: 13-14).

This is a point we shall return to later, but I would first like to consider briefly certain ethical principles and problems of medicine in traditional Chinese and contemporary Western settings. Such a comparison will not deal with differences in ethical perspectives within each setting, but will concentrate on several broadly corresponding ethical principles and issues that appear to be important in both contexts.

Beneficence, defined here in broad terms as a duty to promote the welfare of others (Beauchamp and Childress 1983), is a primary ethical principle of Western biomedicine, with roots that go back to the Hippocratic Oath (Veatch 1981). Veatch rejects the notion that there can be a single overriding principle that will provide a moral framework for all the ethical dilemmas in medicine. However, he states that the tenet that comes closest is that of beneficence, in the sense of trying to produce as much good as possible (Veatch 1981).

Indicating the strong influence of Confucianism, sources on medical ethics in Imperial China emphasized the importance of such Confucian values as humaneness (*jen*) and compassion (*tz'u*) (Unschuld 1978). In this respect, their views expressed characteristics of what in the Western literature is alluded to as an ethics of virtue, with virtue, as applied to practitioners, referring to their moral character, habits and dispositions, rather than the rights, duties and rules that are part of their moral environment (Jonsen and Hellegers 1974; Beauchamp and Childress 1983; Pellegrino 1985). However Beauchamp (1985: 311), following Aristotle, states that virtue is tied to *actions* that *ought* to be performed, and in Chinese ethical statements quoted below, attributes of virtue are combined with specifications of duties or obligations that approximate Western notions of beneficence in emphasisizing actions that will serve the patient's welfare and maximize good. The following examples are from the literature of prerepublican China, as translated and analyzed by Unschuld (1979).

Sun Szu-miao (A.D. 581?-682) was an independent physician who wrote major works on Chinese medicine. He described the guiding principles of what he called a "Great Physician," who must develop first "a marked attitude of compassion," and "who should look upon those who come to grief as if he himself had been struck.... Neither dangerous mountain passes nor the time of day, neither weather conditions nor hunger, thirst or fatigue should keep him from helping wholeheartedly" (Unschuld 1979: 30). Chang Kao (fl. A.D. 1210), a Confucian scholar physician, wrote: "Physicians should remember: 'When another person is ill, it is as though I myself (am ill).' When a physician is called for relief he should respond speedily and without delay" (Unschuld 1979: 52). Kung T'ing-hsien (fl. A.D. 1615), another Confucian physician, urged physicians to "adopt a disposition of humaneness," and he wrote that they "should make a very special effort to assist the people and to perform far-reaching good deeds" (Unschuld 1979: 71).

In the statements of the Chinese writers quoted above, and in the Western medical principle of beneficence, the emphasis is on the physician's duty to help others, his moral obligation to promote the welfare of the sick. An emphasis on personal gain that the physician derives from his practice is perceived as a contrary value. Tension between these values was an ethical problem in the medicine of Imperial China as it is in Western biomedicine.

The Chinese writers quoted above advocated commitment to patients' welfare and admonished against the exploitation of medicine for material rewards. Sun Szu-miao stated that physicians should not "strive with their whole heart for material goods" (Unschuld 1979: 32). One of the central problems considered by Chang Kao in his discussion of ethics was greed versus unselfish help, and he specifically mentioned physicians whose behavior exemplified both (Unschuld 1979: 44 ff.). The relationship between medical ethics and medical costs was treated more ambiguously by Kung T'ing-hsien. On the one hand, he advised patients not to fear expenses, and he asked "what is more precious, life or material goods?" On the other hand, he counselled physicians "not to esteem profits too highly, but instead cultivate humaneness and righteousness" (Unschuld 1979: 72-73). Centuries later, Hsu Yen-tso (fl. A.D. 1895), who wrote treatises on medicine, saw the relationship between mercenary and beneficent purposes in medicine in the following terms: "The intentions of physicians are twofold. One consists in preserving human life, the other consists in making profit. Should we not be cautious in view of these contrary tendencies?" (Unschuld 1979: 110).

Turning to Western biomedicine, in the United States, where doctors have among the highest incomes of any professional groups, the issue of physicians' income as a factor in rising medical costs has been of major public and government concern (Relman 1987). The tension between self-interest and altruism has been defined as the central paradox in medicine (Pellegrino 1985). Ethical problems that reflect this paradox range from broad questions concerning increasing entrepreneurship of physicians fostered by the development of the for-profit "new medical and industrial complex" (Relman 1987) to such individual medical decisions by physicians, as whether to order an additional diagnostic test. Leaving genuine ambiguities of medical judgment aside, a profit-oriented physician might order the test because of addititional income it might provide; a physician whose dominant interest is in the welfare of the patient might refrain from the added risk of a test (Graber, Beasley and Eaddy 1985).

The relationship between medical ethics and professionalization suggests further resemblances between ethical aspects of medicine in Imperial China and the West. Unschuld (1979: 27-28) finds that in China the first indications of group consciousness appear in the writings of Sun Szu-miao, referred to earlier, when he criticized the habit of belittling other physicians; Unschuld points out that such behavior is a decided disadvantage for a group in the process of pursuing professionalization. In this connection, the relationship between avoidance of public criticism of colleagues and the collective interests of the group draws the attention of Freidson (1970) in his discussion of Western professional medicine. As he views it, physicians try to preserve a united front against criticism by outsiders. "If one practitioner cannot restrain himself from criticizing another, he should at least do so in private to the man's face, or at most within a closed professional circle" (Freidson 1970: 179). Both Unschuld and Freidson see codes of medical ethics in China and the West respectively, as efforts to support professional control by assuring the public that practitioners will use medical resources in morally responsible ways.

In Unschuld's opinion, many students of Chinese civilization exaggerate its singularity; he himself calls attention to correspondences he sees between the course of professionalization and medical ethics in China and in the West. He writes, "It was not the results which were at all times the same; the similarity lies in the dimensions of the ethics which were recognized and conceptualized by the core group of independently practicing physicians" (Unschuld 1978: 118).

CONCLUSION

This paper explores some of the ways in which anthropology may contribute to the study of medical ethics, a subject which hitherto has received relatively little attention from medical anthropologists. Much of the current literature on medical ethics pertains to Western biomedicine, with moral issues involving physicians and other biomedical professionals as central matters of concern. Anthropologists can help broaden the perspective for the study of medical ethics by bringing to it a comparative approach which examines in different cultural settings the ethics of folk and popular health care sectors, as well as of professional medicine, including biomedicine.

Examples of such a comparative approach are briefly described in the paper, and the importance of increased knowledge of the ethnoethics of medicine in non-Western societies is emphasized. Medical ethnoethics here refers to moral tenets and problems of health care as they are conceived and reacted to by members of a society.

Although anthropologists in their studies of a wide range of medical systems have given relatively little attention to medical ethics, this does not mean that they have been largely unconcerned about the relationship between morality and medical beliefs and practices in such systems. For example, in many studies of medical systems by anthropologists, indigenous moral values have had an important place in the description and analysis, but primarily as they relate to etiologies that attribute illnesses to social or religious transgressions. In other words, it is as cognitive factors of medical significance rather than as ethical influences on health care that moral values have been mainly viewed in such studies. Yet as an aspect of culture concerned with moral obligations of health care, and in a larger sense to moral aspects of procreation and the duration and quality of life, medical ethics would seem to warrant more attention from medical anthropologists than it has received.

In indicating something of the range and types of problems that might fall within the domain of comparative studies of medical ethics by anthropologists, this paper considers cross-cultural variations and similarities in moral principles and moral issues related to the preservation and restoration of health as well as some social and cultural factors that might help explain these contrasts and resemblances.

It should be emphasized that this paper is not offered as a programmatic statement to orient further anthropological work on medical ethics. Rather, it is intended to be illustrative of the kinds of contributions which anthropologists are capable of making. Stated in broad terms, the objective of the comparative perspective stressed in the paper is better comprehension of the scope, nature and bases of variations and similarities in medical ethics as they are expressed in different cultures.

ACKNOWLEDGMENTS

I would like to thank Jack Bilmes, George Grace, Alan Howard, George Weisz and an anonymous reviewer for their helpful comments on earlier versions of this paper.

REFERENCES

Alexander, Linda
   1979   Clinical Anthropology: Morals and Methods. Medical Anthropology 3: 61-107.
Baer, Hans A., Merrill Singer and John H. Johnsen
   1986   Toward a Critical Medical Anthropology. Social Science and Medicine 23: 95-98.
Beals, Alan R.
   1976   Strategies of Resort to Curers in South India. In Charles Leslie (ed.) Asian Medical
          Systems. Berkeley: University of California Press. Pp. 184-200.
Beauchamp, Tom L.
   1985   What's So Special about the Virtues? In Earl L. Shelp (ed.) Virtue and Medicine:
          Explorations in the Character of Medicine. Dordrecht: D. Reidel. Pp. 307-27.
Beauchamp, Tom L. and James F. Childress
   1983   Principles of Biomedical Ethics (2nd ed.). New York: Oxford University Press.
Bosk, Charles
   1979   Forgive and Remember: Managing Medical Failure. Chicago: University of Chicago Press.
Brody, Howard
   1976   Ethical Decisions in Medicine. Boston: Little, Brown and Company.
Bulger, Roger J.
   1987   The Modern Context for a Healing Profession. In Roger J. Bulger (ed.) In Search of the
          Modern Hippocrates. Iowa City: University of Iowa Press. Pp. 3-8.
Cain, Mead T.
   1977   Village Fertility Study Report No. 4. Economic Class, Economic Mobility and the
          Development Cycle of Households: A Case Study in Rural Bangladesh. Bangladesh Inst. of
          Development Studies.
Cassell, Joan
   1987   On Control, Certitude and the "Paranoia" of Surgeons. Culture, Medicine and Psychiatry ll:
          229-49.
Caudill, William
   1953   Applied Anthropology in Medicine. In A.L. Kroeber (ed.) Anthropology Today. Chicago:
          University of Chicago Press. Pp. 771-806.
Chrisman, Noel J. and Thomas W. Maretzki (eds.)
   1982   Clinically Applied Anthropology. Dordrecht: D. Reidel.
Colson, Anthony C. and Karen E. Selby
   1974   Medical Anthropology. In Bernard J. Siegel, et al. (cds.) Reviews in Anthropology. Palo
          Alto: Annual Reviews, Inc. Pp. 245-62.
Eidelman, Arthur I.
   1986   In Israel People Look to Two Messengers of God. Hastings Center Report 16, 4: 18-19.
Engelhardt, H. Tristam, Jr.
   1986   The Foundations of Bioethics. New York: Oxford University Press.
Fabrega, Horacio, Jr.
   1972   Medical Anthropology. In Bernard J. Siegel (ed.) Biennial Review of Anthropology.
          Stanford: Stanford University Press. Pp. 167-229.
Fabrega, Horacio, Jr. and Daniel B. Silver
   1973   Illness and Shamanistic Curing in Zinacantan. Stanford: Stanford University Press.
Foster, George M. and Barbara Gallatin Anderson
   1978   Medical Anthropology. New York: John Wiley.
Freidson, Eliot
   1970   Profession of Medicine: A Study of the Sociology of Applied Knowledge. New York:
          Harper and Row.
Gaines, Atwood D. and Robert A. Hahn (eds.)
   1982   Physicians of Western Medicine: Five Cultural Studies. Culture, Medicine and Psychiatry
          6, 3.

Gallo, Pia Grassivaro
    1986   Views of Future Health Workers in Somalia on Female Circumcision. Medical Anthropology Quarterly 17: 71-3.
Gordon, Deborah R.
    1988   Tenacious Assumptions in Western Medicine. *In* Margaret Lock and Deborah Gordon (eds.) Biomedicine Examined. Dordrecht: Kluwer Academic Publishers. Pp. 19-56.
Graber, Glenn C., Alfred D. Beasley and John A. Eaddy
    1985   Ethical Analysis of Clinical Medicine: A Guide to Self Evaluation. Baltimore: Urban and Schwarzenberg.
Gruenbaum, Ellen
    1982   The Movement Against Clitoridectomy and Infibulation in Sudan: Public Health Policy and the Women's Movement. Medical Anthropology Newsletter 13: 4-12.
Hahn, Robert A. and Atwood D. Gaines (eds.)
    1985   Physicians of Western Medicine. Dordrecht: D. Reidel.
Hahn, Robert A. and Arthur Kleinman
    1983   Biomedical Practice and Anthropological Theory. *In* Bernard J. Siegel et al. (eds.) Annual Review of Anthropology. Palo Alto: Annual Reviews, Inc. Pp. 305-33.
Harwood, Alan (ed.)
    1981   Ethnicity and Medical Care. Cambridge, Mass.: Harvard University Press.
Hatch, Elvin
    1983   Culture and Morality: The Relativity of Values in Anthropology. New York: Columbia University Press.
Janzen, John M.
    1978   The Quest for Therapy in Lower Zaire. Berkeley: University of California Press.
Jonsen, Albert R. and Andre E. Hellegers
    1974   Conceptual Foundations for an Ethics of Medical Care. *In* Lawrence R. Tancredi (ed.) Ethics of Health Care. Washington, D.C.: National Academy of Sciences. Pp. 3-20.
Kass, Leon R.
    1985   Toward a More Natural Science. New York: Free Press.
Katz, Jay
    1984   The Silent World of Doctor and Patient. New York: Free Press.
Katz, Pearl
    1985   How Surgeons Make Decisions. *In* Robert A. Hahn and Atwood D. Gaines (eds.) Physicians of Western Medicine. Dordrecht: D. Reidel. Pp. 155-75.
Kleinman, Arthur
    1980   Patients and Healers in the Context of Culture. Berkeley: University of California Press.
Koenig, Barbara A.
    1988   The Technological Imperative in Medical Practice: The Social Creation of a "Routine" Treatment. *In* Margaret Lock and Deborah Gordon (eds.) Biomedicine Examined. Dordrecht: Kluwer Academic Publishers. Pp. 465-96.
Koenig, Michael A. and Stan D'Souza
    1986   Sex Differences in Childhood Mortality in Rural Bangladesh. Social Science and Medicine 22: 15-22.
Koo, Jason and Donald O. Cowgill
    1986   Health Care of the Aged in Korea. Social Science and Medicine 23: 1347-52.
Korbin, Jill E. (ed.)
    1981   Child Abuse and Neglect: Cross-Cultural Perspectives. Berkeley: University of California Press.
Kunstadter, Peter
    1980   Medical Ethics in Cross-Cutural and Multi-Cultural Perspective. Social Science and Medicine 14B: 289-96.

Landy, David
    1983   Medical Anthropology: A Critical Appraisal. *In* Julio L. Ruffini (ed.) Advances in Medical
            Social Science. Vol. 1. New York: Gordon and Breach. Pp. 185-314.
Lieban, Richard W.
    1967   Cebuano Sorcery. Berkeley: University of California Press.
    1973   Medical Anthropology. *In* John J. Honigmann (ed.) Handbook of Social and Cultural
            Anthropology. Chicago: Rand McNally. Pp. 1031-72.
Lock, Margaret
    1988   Introduction. *In* Margaret Lock and Deborah Gordon (eds.) Biomedicine Examined.
            Dordrecht: Kluwer Academic Publishers. Pp. 3-10.
Pellegrino, Edmund D.
    1985   The Virtuous Physician and the Ethics of Medicine. *In* Earl E. Shelp (ed.) Virtue and
            Medicine. Dordrecht: D. Reidel. Pp. 237-55.
Pelto, Pertti J. and Gretel H. Pelto
    1973   Ethnography: The Fieldwork Enterprise. *In* John J. Honigmann (ed.) Handbook of Social
            and Cultural Anthropology. Chicago: Rand McNally. Pp. 241-88.
Polgar, Steve N.
    1962   Health and Human Behavior: Areas of Special Interest to the Social and Medical Sciences.
            Current Anthropology 3: 159-205.
Ramanamma, A. and Usha Bambawale
    1980   The Mania for Sons: An Analysis of Social Values in South Asia. Social Science and
            Medicine 14B: 107-10.
Relman, Arnold S.
    1987   The Future of Medical Practice. *In* Roger J. Bulger (ed.) In Search of the Modern
            Hippocrates. Iowa City: University of Iowa Press. Pp. 197-211.
Scheper-Hughes, Nancy (ed.)
    1987   Child Survival: Anthropological Perspectives on the Treatment and Maltreatment of
            Children. Dordrecht: D. Reidel.
Scotch, Norman A.
    1963   Medical Anthropology. *In* Bernard J. Siegel (ed.) Biennial Review of Anthropology.
            Stanford: Stanford University Press. Pp. 30-68.
Spiro, Melford E.
    1986   Cultural Relativism and the Future of Anthropology. Cultural Anthropology 1: 259-86.
Subramanian, K.N. Siva
    1986   In India, Nepal, and Sri Lanka, Quality of Life Weighs Heavily. Hastings Center Report 16,
            4: 20-22.
Turner, Victor W.
    1964   An Ndembu Doctor in Practice. *In* Ari Kiev (ed.) Magic, Faith and Healing. New York:
            Free Press. Pp. 230-63.
Unschuld, Paul. U.
    1978   Confucianism. *In* Warren T. Reich (ed.) Encyclopedia of Bioethics. Vol. 1. New York: Free
            Press. Pp. 200-04.
    1979   Medical Ethics in Imperial China: A Study in Historical Anthropology. Berkeley: Uni-
            versity of California Press.
Veatch, Robert M.
    1978   Medical Ethics Anthologies: Alternatives for Teaching. Hastings Center Report 8, 3: 14-16.
    1981   A Theory of Medical Ethics. New York: Basic Books.
Young, Allan
    1982   The Anthropologies of Illness and Sickness. *In* Bernard J. Siegel et al. (eds.) Annual
            Review of Anthropology. Palo Alto: Annual Reviews, Inc. Pp. 257-85.

MORALITY AND THE SOCIAL SCIENCES

What can the social sciences contribute to morality? An answer to that question depends, not surprisingly, on what morality is taken to be. According to the prevailing positivist approach in Anglo-American philosophy,[1] morality consists of rules and principles, which, because they are normative, can be articulated and defended only on the basis of rational arguments directed at what **ought** to be the case. Because the empirical research of social scientists is directed at what **is** the case, it is irrelevant to establishing the rules and principles constitutive of morality. In positivist morality, therefore, social scientists are consigned to the menial task of discovering facts that can be used in the application of antecedently existing moral standards.

So if social scientists are to be more than handmaidens to philosophers in the moral enterprise, the positivist concept of morality must be jettisoned. This paper presents an argument to that effect. In the next section the positivist concept of morality is elucidated and criticized. Then an alternative contextualist approach to morality is sketched, and two examples of social science research that exemplify a contextualist understanding of morality are provided.

POSITIVIST MORALITY

Positivist morality is about justification. Justification occurs at two levels the level of moral rules and principles and the level of moral judgments that flow from these rules and principles. In both instances justification means rational argumentation. For the justification of rules and principles, there is no canonical form of rational argument. Kant's defense of the categorical imperative in the *Groundwork of the Metaphysics of Morals* and Mill's "proof" of the principle of utility in *Utilitarianism* typify the kinds of arguments that are expected, however. For moral judgments, on the other hand, there is a canonical form of rational argument – justification by subsumption. Particular moral judgments are justified by subsuming facts under rules or principles to generate decisions. This deductive model of moral justification is the counterpart of the deductive-nomological model of explanation in positivist philosophy of science.

Implicit in this account of justification is the assumption that the content of morality is propositional. Morality can be exhaustively expressed (with the qualification, of course, that this is what is possible *in principle*) in a set of explicit propositions about what ought to be the case along with the conclusions that can be deduced from the conjunction of these normative propositions and empirical propositions. The goal of moral theory, the subject of moral philosophy, is to construct a comprehensive and rigorous, yet simple and elegant system of rules and principles from which moral judgments can be derived. Morality, in

241

G. Weisz (ed.), Social Science Perspectives on Medical Ethics, 241–260.
© 1990 *Kluwer Academic Publishers. Printed in the Netherlands.*

positivist terms, consequently becomes equated with moral theory, so that the better moral decision maker turns out to be the better moral philosopher.[2]

What is wrong with the positivist concept of morality? Because the positivist approach reduces morality to moral theory, that question can be answered by seeing what is wrong with moral theory. In 1912 Prichard, in his famous article, "Does Moral Philosophy Rest on a Mistake?" expressed misgivings about moral philosophy which he attributed, in part, to the "increasingly obscure" aim of the subject. Prichard posed three questions that continue to challenge those working in the area (Prichard 1912: 21): (1) "What are we really going to learn by Moral Philosophy?" (2) "What are books on Moral Philosophy really trying to show, and when their aim is clear, why are they so unconvincing and artificial?" and (3) "Why is it so difficult to substitute anything better?" These questions can be summed up by asking, "What is moral theory for?"

Three functions for moral theory are commonly adduced. First, moral theory is action-guiding. In positivist morality the conclusions of deductive arguments from rules or principles are supposed to be determinate propositions about what ought or ought not to be done. Second, moral theory provides criteria for determining what features of the world are morally relevant. How does one distinguish propositions that are morally relevant, for example, her sister was allowed to get her ears pierced when she was eight years old, from propositions that are not morally relevant, for example, she was born in March? In positivist morality the terms and concepts in the rules and principles under which facts are to be subsumed supposedly determine what facts are pertinent. Third, moral theory is the source of normativity. Because of an allegedly insuperable gulf between facts and values, the realm of normativity – of "oughtness" – is segregated from the realm of descriptivity – of "isness" – and thus must be accounted for in its own terms. In positivist morality the rules and principles constitutive of theory are regarded as normative in the sense that their rational justifications make them binding on all rational creatures. A critique of positivist morality, then, amounts to showing that moral theory cannot fulfill these three functions.

Why can't moral theory be action-guiding? For the familiar reason that the incongruity between the abstract terms of moral rules and principles and the particular descriptions of facts prevents the subsumption of facts, and thus the deduction of moral conclusions, from proceeding in a straightforward fashion. There is a crucial "application gap" between general norms and specific facts that can be bridged only by considerations external to moral theory. To take an example from medical ethics, how is the principle, "Doctors must respect the autonomous decisions of patients," to be applied in a situation in which a thirty-eight-year-old man with a mild upper respiratory infection, severe headache, stiff neck, and high fever has his complaints diagnosed in an emergency room as pneumococcal meningitis and then refuses to consent to treatment because he wants to be allowed to die (Miller 1981: 22)?[3] This bacterial meningitis is almost always fatal if not treated, and if treatment is postponed, permanent neurological damage is likely. Does "autonomy" mean that this patient's decision must be respected? Does, in other words, the following deduction go through here?

1) Doctors must respect the autonomous decision of a patient.

2) This patient's decision is autonomous.

3) Therefore, doctors must respect this patient's decision.

What obviously is both crucial and controversial is the second premise. Is merely being intentional and free (in the sense of not externally coerced) sufficient to make a decision autonomous? If so, two supplementary premises are needed for the deduction to be valid:

1*) An autonomous decision is one that is intentional and free.

1**) This patient's decision is intentional and free.

And if not, how is the notion of autonomy to be understood in medicine? Can a univocal account of autonomy be formulated, or must the concept of autonomy be tailored to fit diverse medical contexts?

The problem here, endemic to all general norms whether moral or legal, is one of classification. Does this particular decision fall within the ambit of the general term "autonomous"? Any attempt to apply rules and principles faces classification problems, which cannot be resolved by these standards themselves. Rather, extrinsic considerations must be introduced to decide whether a particular fact situation falls within the compass of a general term. Because these considerations are extra-theoretic, the application gap shows that the positivist concept of morality is incomplete.

It might be objected that this criticism appears plausible only because moral theory has been defined in too niggardly a fashion. A moral theory should be taken to include not only normative rules and principles but also definitions of the terms contained in these standards. One should adopt, in other words, a more capacious model of theory similar to that used in logic or mathematics. But this expansion of theory to include missing considerations only postpones the problem by assuming that complete, explicit definitions of general moral terms are possible. It views the problem of vagueness as intrinsic to language and thus resolvable, at least in principle, by rigorous conceptual analysis. If only we thought hard enough about these terms, this objection presupposes, we could produce explicit sets of necessary and sufficient conditions for their application.

Vagueness is not, however, an intrinsic defect of language; rather it results when language bumps up against the world. One cannot, that is, classify terms as vague or not-vague on *a priori* grounds. Any term, even one that appears specific, fixed, and determinate, can become vague if the world kicks up the right concatenation of facts in the right context. Take, for instance, the notion of chicken. We all know the difference between chicken and beef, and we all know the difference between chicken and turkey. How can "chicken" be vague? But a case in contract law had to answer the question, "What is a chicken?"[4] The plaintiff company contracted with the defendant company to buy a fixed number of pounds of chicken at an agreed price. The plaintiff expected young fryers and roasters but received old fowl and stewing chickens and thus sued for breach of contract. The court had to decide the meaning of "chicken" for the purposes of this contract. Legal commentators, notably H.L.A. Hart (1961: 125), well understand the impossibility of anticipating all possible scenarios and formulating standards in advance for dealing with them. The application of moral as well as legal norms inevitably requires an appeal to considerations beyond the bare norms.

One response to the application gap is to introduce intermediate or mid-level norms (Bayles 1984, 1986). In medical ethics, for example, a principle of informed consent might be used to try to span the gap between a general principle of autonomy and particular decisions that have to be made about consent to medical treatment or participation in medical research. A mid-level, more specific principle of informed consent does appear to be more action-guiding or directive because it is closer to the description of the problem that needs to be resolved. The difficulty now, though, is explaining the relationship between the mid-level principle of informed consent and the higher-order principle of autonomy. Presumably the principle of autonomy "justifies," in some sense, the principle of informed consent. But what exactly does "justification" mean here?[5] Presumably a deductive argument in which the principle of autonomy figures as the initial premise and the principle of informed consent as the conclusion. But does "justification" require that there be only one such deductive argument, namely, the one with the principle of autonomy as the first premise? What if similar arguments could be constructed (and they easily could) with, say, a principle of utility or a principle about rights as the first premise? If the principle of informed consent could be "justified" in terms of several higher-order principles, then it would not be "justified" in terms of any of them – these higher-order principles would be idle. But how could such multiple "justifications" be excluded? Only by showing which of these higher-order principles is the "correct" moral principle. That would require an accepted method for adjudicating the competing claims of Kantianism, utilitarianism, and rights-based theories, but no such method exists. Thus introducing intermediate or mid-level principles creates as big a problem as the one it putatively solves.

It might be objected, finally, that the application gap exists for only hard or controversial cases. In the overwhelming majority of cases, the world is not recalcitrant, so the determinacy of moral rules or principles allows their application to proceed smoothly. In the comparatively rare instances in which their application is not straightforward, moral theory may "run out," but that is insufficient reason for jettisoning it. This overall picture of morality is, I think, correct. Morality is largely settled. Morality encompasses the commonplace as well as the controversial, and it is only a preoccupation with the latter that creates the unwarranted impression that moral issues are, by their very nature, disputatious.

Why is this admission not damaging? Because it equivocates on the notion of morality. Rules of moral theory are not rules of moral practice. Quotidian morality is settled because the rules of the practice of morality are settled. When corresponding rules of theory are created by philosophers, they simply overlay antecedently existing rules of practice and therefore are redundant (Walzer 1985: 6; Noble 1979: 502-503). Moral issues become controversial when the rules of quotidian morality cannot handle them. At that point rules of theory can be invoked but succumb to the application gap. Thus, the first indictment of positivist morality does need to be amended. Positivist morality is not action-guiding because it is unnecessary in the large class of prosaic moral cases and unhelpful in the small class of controversial moral cases.

What about the claim that moral theory provides criteria for determining which features of the world are morally relevant? Its failings in this respect can be exposed by examining the two main pretenders to the crown of moral theory, Kantianism and utilitarianism. The leading defender of a practically helpful version of Kantian ethics is Onora O'Neill (formerly Onora Nell). In her book, *Acting on Principle* (Nell 1975), she contends that Kant's categorical imperative is useful in guiding action in two respects. It can, on the one hand, operate internally or subjectively to tell an agent whether a contemplated action possesses moral worth; it can also operate externally or objectively to tell anyone, agent or observer, whether an action is morally obligatory, forbidden, or permissible. In Kant's view these are distinct issues. The performance of a morally obligatory or permissible action does not possess moral worth unless the agent also acts from the proper motive, namely, the motive of duty. Telling the truth because one fears being found out, for example, is morally correct but lacks moral worth. Telling the truth has moral worth only when it is done because it is the morally right thing to do.

But how is the categorical imperative to be applied? Kant's first formulation of the categorical imperative is: "Act only according to that maxim whereby you can at the same time will that it should become a universal law." For Kant the moral status of an action is a function of that action's maxim. There are, accordingly, two steps in the application of the categorical imperative: (1) formulating the maxim of an action; and (2) testing this maxim for consistency, that is, seeing whether this maxim could hold for all moral agents (and thus be a "universal law"). The problem of determining the morally relevant features of the world infects the first step. How are maxims to be stated? A parent's different treatment of two children, for instance, could be described as favoritism or as an attempt to reward initiative. A test for consistency is likely to produce different outcomes, depending upon which description is selected. (If you were the disadvantaged child, would you countenance favoritism on the part of your parent? What about an even-handed parental policy of rewarding initiative?) But which is the morally relevant description?

O'Neill's initial answer is that a maxim is an agent's subjective intention, so the specification of a maxim depends upon the agent's sincere and honest report of what he or she intends to do. As O'Neill recognizes, however, this introduces all sorts of problems. Maxims can be infected by non-culpable as well as culpable ignorance, bias, and self-deception, and for even a sincere agent there can be competing plausible accounts of how a maxim should be formulated. O'Neill's conclusion is not encouraging:

> if we are to find a way of extending Kant's theory of right to contexts of assessment [i.e., to determining whether actions are obligatory, forbidden, or permissible], some further solution to the problem of relevant descriptions is still needed.... Any adequate solution must state an effective criterion of selection that is generally applicable and yields plausible results. I cannot offer any solution which meets these standards. Kant's theory of right can at least be used in contexts of decision and action [i.e., to determine whether an action has moral worth] (Nell 1975: 142-143).

The intractability of this problem subsequently forces O'Neill to abandon the claim that the categorical imperative can be used to determine whether actions are obligatory, forbidden, or permissible (O'Neill 1985). Kant's categorical imperative, in other words, cannot tell us whether an action is morally right or wrong because there is no acceptable account of moral relevance. The most the categorical imperative can do is tell an agent whether a proposed action possesses moral worth, but success in even this regard presupposes that the agent is honest, sincere, and cognizant of subterranean motives.

One might think that, given David Lyons' argument that utilitarian generalization and certain forms of rule utilitarianism collapse into act utilitarianism (Lyons 1965),[6] utilitarianism fares better in this respect, because Lyons' argument, after all, turns on what counts as a morally relevant description of an action for a utilitarian. His view that all the causal consequences of an action are relevant for utilitarianism is a purely formal account of relevance, but it is all he needs for his argument. If all the causal consequences of an action are morally relevant, then, given a common theory of value or utility for assessing those consequences, the morally relevant features of any action will be the same for utilitarian generalization and act utilitarianism, and once this assimilation has occurred, Lyons can convincingly argue that there is no difference between the outcomes of utilitarian generalization and act utilitarianism.

But such a formal account of moral relevance is inadequate for substantive issues. Consider, for example, the problem of determining for what persons are responsible, and recall Holmes' view that in the law an action is simply a voluntary muscular contraction and everything that causally follows is a consequence of that action (Holmes 1881: 91-95). Thus for Holmes when I raise my arm with a loaded gun in my hand, point it at you, and pull the trigger, my action consists simply of the voluntary muscular contractions in my arm, hand, and fingers, and everything else is a consequence. But this view, as Holmes realized, creates problems for the determination of liability because unintentional and unforeseeable consequences are causal consequences of actions, yet persons sometimes are not held responsible, legally or morally, for them, particularly when they are remote or bizarre. The inadequacy of a formal solution to this problem is demonstrated by the doctrinal incoherence surrounding the notions of "but for" and "proximate" causation in tort law.[7]

Another difficulty for utilitarianism is how to interpret the claim that the consequences for all sentient creatures are morally relevant. What counts as a sentient creature? Animals, clearly, but then for how much do animals count? That is a central issue in the debate over the moral permissibility of medical research on animals. Proponents of animal rights want animals to count equally with human beings, while defenders of medical research claim that animals should not be accorded the same moral protection as human beings. The standard utilitarian tack here is that because animals cannot have certain cognitive and emotional experiences, because they cannot remember, form expectations, and feel remorse and regret, for instance, they do not merit as much moral consideration as persons. But then what about fetuses, those other important entities on the moral margin? They likewise lack the capacity to remember and anticipate, to feel guilt and shame, so do they count for the same as animals, or does their potential to have these experiences elevate them morally, and if so, how much? How does a utilitarian decide when the criterion of moral relevance is the capacity to experience utility, however "utility" is defined, and when it is the potential for this capacity?

A similar conclusion follows from this second indictment. Positivist moral theory remains unproblematic as long as it simply incorporates, overtly or covertly, the criteria of moral relevance that suffuse the ordinary practice of morality. When controversies arise, however, it becomes obvious that positivist morality possesses no workable account of moral relevance. Solutions to these difficulties must be sought beyond its narrow boundaries.[8]

Can positivist morality nevertheless account for normativity? That is a hard question to answer because the notion of normativity is vague. In one sense moral theory might be thought to be normative in virtue of its action-guidingness. If moral theory could prescribe and proscribe actions, it would *ipso facto* be normative, but, as we have seen, moral theory is dispensable in "easy" cases and impotent in "hard" cases. In what other respects might moral theory be normative, then?

The best way to proceed, again, is by examining actual moral theories, in this case Rawls' early notion of moral theory as explication (Rawls 1951). The aim of moral theory here is to formulate a set of principles that will explicate the considered moral judgments of competent judges. Once this set of considered judgments has been identified, the task is to construct a comprehensive, yet simple and elegant, set of principles that explicates these considered judgments. A considered judgment has the form: "Since a, b, and c ... are the facts of the case, and m, n, and o ... are the interests that conflict in the case, m is to be given preference over n, o ..." Considered judgments simply report decisions about cases; they do not contain reasons for those decisions. Because it would make no sense to have a separate principle for each case, the principles must be abstracted to some degree from the particular facts and interests enumerated in the

considered judgments. After the set of principles is constructed, it is tested by seeing whether a competent person, explicitly and consciously applying these principles in an intelligent and consistent way to the same cases from which they were
derived, would arrive at exactly the same considered judgments for those cases. If
the set of considered judgments generated in this way by the principles is identical with the set of considered judgments that generated the principles, then the
principles constitute a satisfactory explication of those considered judgments.

All of this looks entirely descriptive, however, and that impression is
reinforced by Rawls' comment that ethics should be regarded as analogous to
inductive logic. Where does normativity enter? There are, I think, three possibilities. The first is that it comes from a "sanitized" set of considered judgments.
Producing a set of considered judgments is a group effort, so that in formulating
this set, individual biases, predilections, and idiosyncrasies are likely to cancel
out. Agreement among a group of competent judges makes it plausible to claim
that these considered judgments are "correct," and thus that the principles that
reflect them are the "correct" moral principles. But if this is what normativity
amounts to, normativity does not come from theory, but rather from social
agreement or consensus about the starting-points for the construction of theory.
Formulating principles is simply a way of making the content of considered
judgments more perspicuous. In what sense theory itself could be normative
therefore remains mysterious.

The second possibility is that theory itself is normative because its principles
can be applied to new problems to resolve the moral perplexities they raise. Just
as a scientific theory can be extended beyond the data upon which it rests to
generate predictions about new phenomena, so a moral theory can be extended
beyond the considered judgments upon which it rests to generate decisions for
new cases. But how does one know which principle is the correct principle to
apply to a new case? Because principles must abstract from considered judgments, the facts and interests mentioned in principles are unlikely to match
exactly the facts and interests in the description of a problematic case. Moreover,
given this slippage, more than one principle may be applicable to a new case, and
these principles may conflict. The difficulties in applying rules or principles to
cases surface once again. These difficulties are, in fact, theoretically insoluble, as
Rawls' response to this problem reveals. Rawls says that a principle must yield a
result which "after criticism and discussion, seems to be acceptable to all, or
nearly all, competent judges, and to conform to their intuitive notion of a
reasonable decision" (Rawls 1951: 188). Thus it is considered judgments, not
principles, that continue to do the work. The judgments of individuals about
specific cases remain the touchstone of moral correctness.

The third possibility is that theory is normative because principles can bring
about the rejection or revision of considered judgments. Principles would be
doing genuine normative work if they could be used to assess considered
judgments for correctness. Can they? Do principles ever conflict with considered
judgments and prevail over considered judgments in some of these conflicts?

They do not. Rawls talks about the capacity of a principle "to hold its own ... against a subclass of the considered judgments of competent judges," and he says "this fact may be evidenced by our intuitive conviction that the considered judgments are incorrect rather than the principle, when we confront them with the principle" (Rawls 1951: 188). Notice that the basis for rejecting a considered judgment is not a principle, but rather an "intuitive conviction" that the considered judgment is wrong. Principles serve only as *occasions* for reassessing considered judgments, not as *grounds* for such reassessments. A principle, in other words, merely points to the need for a reassessment, it does not show how this reassessment should be conducted or what the outcome should be. This is clear in one of Rawls' subsequent remarks: "while the considered judgments of competent judges are the most likely repository of the working out of men's sense of right and wrong,... they may, nevertheless, contain certain deviations, or confusions, which are *best discovered* by comparing the considered judgments with principles ..." (Rawls 1951: 189; emphasis added).

Rawls' discussion is confusing because he moves back and forth between regarding principles as occasions for reassessments and grounds of reassessments. He initially sees principles as occasioning conflicts which are resolved by intuitive convictions, but then he slips into regarding principles as grounds for resolutions by talking about principles "successfully militat[ing] against" what are "taken to be" considered judgments (Rawls 1951: 189). Having demoted considered judgments to the status of *prima facie* or provisional considered judgments, Rawls feels the need to explain how competent judges could come to hold mistaken considered judgments, and his proposals are familiar. A mistaken considered judgment could result from a mistaken factual belief or bias. Rawls views such a psychological explanation as bolstering the case for rejecting a considered judgment because it conflicts with a principle, but there is no case here to be bolstered. Discovering that a considered judgment rests on a mistaken factual belief or bias is itself reason for rejecting that considered judgment, not ancillary evidence that makes one feel more comfortable about rejecting a considered judgment because it conflicts with a principle.

In sum, Rawls does not explain how moral principles can have normative or justificatory force. His claim that principles exhibit the capacity "to alter what we think to be our considered judgments in cases of conflict" (Rawls 1951: 189) fails to distinguish principles as occasions for reassessments of considered judgments from principles as grounds for reassessments of considered judgments. Once this distinction is drawn and principles are seen as only occasioning reassessments, principles lose their normative force. In this respect as well, moral theory fails to accomplish one of its self-appointed tasks.[9]

Within a positivist concept of morality, the crucial problems of applying moral rules or principles and determining what features of the world are morally relevant can be solved only by appealing to considerations outside of moral theory. These extra-theoretic considerations do the real moral work, and once their important work is finished, justification by subsumption is jejune, just as explanation by subsumption in science is jejune. Moreover, moral theory cannot account for normativity. If normativity is equated with the action-guidingness provided by justification by subsumption, it is illusive; if normativity is something else, it is elusive.

The strategy behind this critique of positivist morality should be appreciated. I do not think it is possible to construct an *a priori* argument to the effect that moral theory is in principle incapable of fulfilling its three putative tasks. The goal must be more modest. This critique amounts to showing that the most sophisticated and representative accounts of moral theory do not fulfill these functions. The strength of the critique rests on two claims: that the objects of the criticisms are exemplars of moral theory and that the gist of the criticism, namely, that moral theory cannot succeed on its own terms, is serious. If these claims are accepted, the upshot is that a positivist model of morality, a model that holds that morality can be exhaustively captured by the systematically ordered, consistent, explicit propositions of rules or principles and that the justification of moral judgments proceeds by subsuming facts under these rules or principles, is bankrupt.

CONTEXTUALIST MORALITY

Contextualist morality is about understanding. Its aim is to explain the practice of morality. Because the preoccupation with justification is abandoned, there is no method of morality, let alone a "rational" method of morality. Moral decision making is, instead, a matter of "muddling through," or coping with, problems as they present themselves.[10] In the absence of a universal method, such as applying the categorical imperative or the act utilitarian principle to every conceivable moral problem, different approaches can be appropriate for different situations. Part of the challenge of moral decision making is figuring out what kind of approach fits the issue at hand.[11] Such an inquiry can be dismissed as non-rational, or perhaps even irrational, only if rationality is restrictively and stipulatively defined as the utilization of a universal method, as it is in the positivist approach.

Contextualist morality rejects the claim that morality can be exhaustively expressed in a set of explicit propositions and thus the identification of morality with moral theory that follows from this claim. The focus of contextualist morality, rather than being theory, is practice. Its concern is accounting for the phenomena of morality. But understanding the practice of morality requires that this practice be located in its social and historical contexts. Morality becomes intelligible only when the background that makes it possible is considered.

Of what does this background consist? One of the most concise and clearest accounts of the "shared background which is the foundation of everyday intelligibility," is Dreyfus' explanation, in the context of a discussion of science, of the three ways in which Heidegger sees explicit understanding as involving "pre-understanding":

> 1. *Vorhabe* (fore-having). The totality of cultural practices ... which "have us" or make us who we are, and thus determine what we find intelligible. In any science this is what

Kuhn calls the "disciplinary matrix" – the skills a student acquires in becoming a scientist, which enable him to determine what are the scientifically relevant facts.

2. *Vorsicht* (fore-sight). The vocabulary or conceptual scheme we bring to any problem. At any given stage in our culture this is captured in a specific theoretical understanding of what counts as real spelled out by the philosophers. In science, *Vorsicht* is whatever is taken to be the relevant dimensions of the problem.

3. *Vorgriff*. A specific hypothesis which, within the overall theory, can be confirmed or disconfirmed by the data (Dreyfus 1980: 9-10).

This tripartite division fits morality, too. The *Vorgriff* would be specific judgments about particular moral issues; the *Vorsicht* the way in which moral issues are conceived and formulated, for instance, as conflicts between individual rights or as raising feminist concerns; and the *Vorhabe* the set of cultural beliefs that delimit the moral realm and thus determine what comes to be identified as a moral issue, for example, the beliefs that explain why *in vitro* fertilization is a hot moral topic but reconstructive microsurgery of Fallopian tubes, which costs approximately $20,000, is not.

Positivist morality founders because it ignores the *Vorhabe* and the *Vorsicht*, where problems of moral relevance and classification arise and must be settled. Contextualist morality thus suggests some easy answers to Prichard's three questions. First, we are going to learn very little, if anything, about the practice of morality, which encompasses the *Vorsicht* and the *Vorhabe* as well as the *Vorgriff*, from moral philosophy. Second, even when books on moral philosophy are clear, they are unconvincing and artificial because they address only one part of morality and pretend it is the whole. Kantians and utilitarians, for example, select a single moral insight from moral practice and elaborate this insight into a sophisticated moral theory. The theory that results, however ingenious and complex, reflects a truncated understanding of what morality is.[12] Finally, it is difficult to substitute anything better because contemporary moral theories are constructed within a positivist framework and therefore ignore moral practice and its background.

The main virtue of contextualist morality is that it can accommodate salient features of morality that are incomprehensible to positivist moral theory. Two developments in medical ethics, for example, are the necessity for anyone working seriously in the area to be familiar with the clinical dimensions of problems, a familiarity that cannot be obtained simply by reading about those problems, and the virtual definition of the field in terms of cases that function as paradigms, for instance, the Karen Ann Quinlan, Joseph Saikewicz, Barney Clark, and Baby Doe cases. The explanation of these two requirements is the same in morality as it is in science, namely, these are ways of acquiring the skills or know-how that constitute the *Vorhabe* of morality. As Dreyfus explains for science:

Like any skills, the practices which make natural science possible involve a kind of know-how (*Vorhabe*) which cannot be captured by strict rules. Polanyi stresses that these skills cannot be learned from textbooks but must be acquired by apprenticeship, and Kuhn adds that they are also acquired by working through exemplary problems (Dreyfus 1980: 16).

The importance of working through problems in medical ethics has even resulted in a call for reviving the tradition of moral casuistry (Jonsen 1986).

In addition, the fundamental nature of some of the moral questions society now confronts, in particular, issues associated with new reproductive technologies and the creation of new life forms by genetic engineering, places them outside the ken of positivist morality. These issues are intractable for positivist morality because they call into question the presuppositions that structure the context within which positivist moral theories operate. Coping with them requires a reconfiguration of the *Vorhabe* and *Vorsicht* of morality, a task for which positivist morality lacks the resources.

The overriding worry about contextualist morality is that it tosses normativity into the trash can along with moral theory. Contextualist morality has ample room for critical normativity, however. Moral practices can be criticized along the lines that Geuss suggests ideologies can be criticized (Geuss 1981: 26-44). A form of consciousness is ideologically false, according to Geuss, if it commits an epistemic, a functional, or a genetic mistake, but these mistakes can also arise in morality.

There are four kinds of epistemic mistakes. One type involves mistakes about epistemic status, for example, thinking that additional scientific evidence could answer the question of when a fetus becomes a person. A second is falsely believing that a social phenomenon is a natural phenomenon, for example, the belief that a woman's place is in the home because women as a matter of fact spend so much time doing housework. A third is falsely believing that the particular interest of a subgroup represents the general interest of a group, for instance, the belief that what is in the interest of doctors is in the interest of society. The fourth is mistaking self-fulfilling beliefs for beliefs that are not self-fulfilling. Two examples are beliefs concerning the capacities of mentally handicapped persons and beliefs about the abilities of patients to handle bad news.

Geuss' notion of functional criticism cannot be directly transposed to moral practices because ideologies are defined as world-pictures that stabilize or legitimize domination or oppression. Functional criticism nevertheless could be pertinent to the practice of morality if it is designed to achieve certain goals within societies, a suggestion that has considerable merit. Hart, for example, holds that law protects persons, property, and promises (Hart 1961: 195), and these could just as easily be goals of morality. Thus, morality could be criticized for not fulfilling its functions, too.

Because positivist philosophers rigidly distinguish justification and explanation, they regard considerations pertaining to the origin, motivation, or causal history of a moral judgment as irrelevant to critical appraisal of that judgment.

But as Geuss points out, there are two senses in which genetic factors are relevant. First, discovering the reasons why one in fact makes a judgment and seeing whether one can acknowledge those reasons can be part of the assessment process. Second, understanding the point of a judgment can depend upon understanding its history. Again, there is a parallel with law. One guide to interpreting statutes involves an appeal to the end, goal, or purpose of a statute, for example, determining the particular defect that enactment of a statute was intended to remedy. A comparable inquiry into the source of a moral judgment could shed light upon the point, and thus the soundness, of that judgment. Genetic considerations presuppose that moral judgments have causal histories and purposes, that they are not timeless and invariant. That presupposition is not satisfied by the judgments that emanate from the pristinely rational rules or principles of positivist moral theories, but it is satisfied by the judgments embedded in moral practices.

These examples of the types of rational criticism to which moral practices are susceptible display two distinctive features. On the one hand, they are directed at moral judgments situated in particular temporal and social contexts. They assume that morality changes and that rational criticism can contribute to that change. On the other hand, these piecemeal criticisms demonstrate the weaker rationality compatible with contextualist morality. Normativity predicated upon rationality has a place in contextualist morality, but the notion of rationality here is more modest and, for that reason, more realistic than the overweening rationality of positivist morality.

TWO EXAMPLES OF CONTEXTUALIST RESEARCH

The social sciences should find contextualist morality hospitable because the goal of understanding morality opens up inquiries into the history of morality as well as moral sociology and anthropology.[13] But although a contextualist approach invites social science research, it also imposes restrictions on the kind of research that should be done. Contextualist research is not compatible with positivist social science.[14] How, then, can the social sciences go about illuminating morality?

A. Genetic Counseling

One way is by studying how moral decisions actually get made. As soon as a universal regulative method for moral decision making is abandoned, the question of how individuals in fact "muddle through" moral problems becomes interesting. Research by Lippman-Hand and Fraser (1979) into how parents make decisions after receiving genetic counseling nicely illustrates the point that moral decision making is essentially a matter of finding an approach that allows one to cope with the nature of the problem. This outcome is ironic because their research was aimed at assessing the adequacy of genetic counseling within a normative framework that defines a rational decision as one that appraises the likely consequences of alternative courses of action and chooses the course of action that leads to the most desirable or least undesirable consequences. The job of a genetic counselor, assuming this consequentialist method of moral decision making, is to provide clients with the empirical information they need to understand the alternatives open to them. Once clients have the various courses of action and the probabilities that attach to the consequences of these options explained to them, they can evaluate the potential outcomes in terms of their own values or utilities and choose the course of action that is most desirable or least undesirable. Thus the effectiveness of genetic counseling depends, given this model, upon whether clients actually make such "rational" consequentialist decisions after receiving genetic counseling.

By interviewing prospective parents Lippman-Hand and Fraser found that clients do not make the "rational" decisions they are supposed to make. The first step they take is to discard the probabilities the counselors have given them. They reduce the alternatives to two equally likely outcomes – either we will have a defective child or we won't. Then they imagine what it would be like to care for and raise a defective child. If they feel they can cope with these scenarios, they try to conceive; if not, they seek permanent contraception. And if they cannot make up their minds, they play "reproductive roulette," that is, they engage in unprotected intercourse, thereby abandoning the decision to fate.

This research shows that an *a priori* method of rational decision making does not fit how decisions are made when parents confront a substantial risk that any child they conceive will be handicapped. Yet the explanation of how parents actually make these decisions rings truer than the prescription that these decisions ought to be made on consequentialist grounds. Parents tailor these momentous decisions to their own needs, abilities, resources, and circumstances. They are creative, active moral agents rather than detached puppets applying impersonal moral norms to their own circumstances. Their individualized coping betrays the presumed universalizability of morality, that is, the view that if a decision is right or wrong for someone, that same decision is right or wrong for anyone else in similar circumstances. Positivist moral theory takes universalizability to be definitive of morality and thus can neither account for nor accept what these parents do. For positivist moral theory this appreciation of how parents in fact make decisions is irrelevant because it does not pertain to the justification of their decisions. Justification can result only from applying a method of morality, and having renounced any such method, these parents can at most be the focus of an interesting study in deviant morality.

B. Reproductive Counseling

Joffe's study of front-line family planning workers traces how three conflicting ideologies of family planning – family planning as a medical service, family planning as an issue of women's rights, and family planning as a profamily issue

– get translated into workable policies (Joffe 1986). As she points out, these three perspectives lead to three different characterizations of the problem of providing family planning services: members of the family planning establishment see the issue as medical; feminists see it as political; and the profamily movement sees it as moral. How a problem is defined can crucially determine how it gets resolved, largely because the definition circumscribes the considerations deemed relevant. Problem definition belongs to the *Vorsicht* of morality, however, so it does not garner the attention of positivist morality.

But Joffe is sensitive to the contextualist dimensions of her project. She observes of feminists, for example, that they are "somewhat belatedly ... coming to the realization that family planning services not only have implications for individual women's 'rights' but also raise difficult questions about the shaping of sexual experience in American culture" (40). Specific claims about rights represent the *Vorgriff* of morality, while locating these claims in the broader context of sexuality is a foray into the *Vorsicht* and *Vorhabe* of morality.

Joffe must, of course, adopt a strategy that fits the nature of her research enterprise, and she does. She rejects a demographic approach in favor of "an ethnographic study of the workplace in the interactionist tradition of Everett Hughes and his associates" (7), because the way in which a demographic perspective emphasizes the outcomes of programs allows "little discussion of the actual meaning of family planning as a social institution" (10).

Joffe cannot shake a positivist approach to social science research entirely, though, because she remains concerned about the "generalizability" of Urban, the family planning clinic she studied (58). In contextualist research the concern is whether an object of study is paradigmatic, not whether it is representative of a broader class of subject-matter. Joffe appears to waffle between these alternatives because although she argues for the latter, she also emphasizes that Urban is a "typical" family planning clinic, a claim that can be interpreted as an attempt to establish its paradigmatic credentials. Joffe does not help matters because at one point she construes "typical" in positivist terms herself, but fortunately in a context that makes it clear that she rejects its positivist meaning: "rather than argue that the women I observed at Urban are 'typical' of family planners everywhere in the United States, I would pose the intended contribution of this study in somewhat different terms" (60). Her final word on the matter firmly establishes the contextualist nature of this research: "We might think ... of the Urban counselors as particularly reflective guides to the general difficulties involved in regulating the sexuality of others" (60).

The substance of Joffe's research shows how a contextualist approach handles issues that bedevil positivist morality. When, for example, is age morally relevant? Positivist morality has no answer to that question, yet age is decidedly relevant for family planning counselors, who find teenage clients particularly difficult. The special circumstances and problems of teenagers require different services. One response to that recognition was that teenagers frequently received a different kind of counseling because "the cardinal rule of contraceptive work –

'The counselor's role is only to facilitate decision making' – was harder to honour with these clients" (78). Rather than attempting to "justify" this different treatment (however that might be done), contextualist morality tries to explain why sexually active teenagers are different and why family planners respond differently, personally and professionally, to the challenges they present.

Joffe also shows how work in an abortion clinic caused counselors to abandon their previous "simpler" views about the morality of abortion in favor of an appreciation of the "more complex reality" of the problem (114-115), and how the moral dilemmas of abortion cannot be separated from the contexts in which they arise (116, 132-134). A crucial issue with respect to abortion concerns the involvement of a woman's family, but, as Joffe points out, "family" is a vague concept (151). Rather than treating this as a classification problem, however, Joffe suggests that "perhaps it would be more relevant to ask how fluid a concept clinics have of 'family' and how far they are willing to go to accommodate their clients' perceptions [of who counts as 'family']" (151). In other words, instead of formulating a definition of "family" and then trying to resolve concrete issues through the application of this definition, one should seek to understand how the problem of who counts as family manifests itself in different circumstances and what considerations lead clinics to adopt the positions they do.

The most important lesson of this study, though, is the need for focusing on concrete cases rather than abstract rules or principles. The family planning counselors understood that there were no simple solutions to the problems they faced and therefore that rigid policies governing their work were impossible. None of the competing ideologies associated with the family planning establishment, the feminist movement, and the profamily movement dominated their choices. Counselors were partially influenced by all three and tailored their decisions to the demands of the moment:

> contrary to a bureaucratic logic that would treat all persons essentially alike, authentic family planning work must speak to the unique circumstances presented by an individual client and her network. In practice, this "contextual" approach means that different clients will be offered different contraceptive methods, with the choice of the method dependent on social as well as medical factors; that counselors will make differential efforts at involving the kin and partners of different clients in the activities of the clinic; and, most significantly, that the ideological messages about sexuality that are transmitted in counseling sessions will also vary according to the circumstances of the client ... (159).

Just as these contextual decisions are incompatible with a "bureaucratic logic," they are likewise incompatible with the rules and principles of positivist morality. General rules and principles are incapable of capturing, let alone "justifying," the complex discriminations and judgments embodied in these decisions.

Joffe's conclusion is an apt summary of the value of her study in particular and of contextualist morality in general:

the most important contribution that Urban counselors and their colleagues have to offer the larger society is a moral standpoint that is nuanced enough to recognize the necessity of accepting people as they are even while one simultaneously struggles to create a better society (166).

The interplay between the descriptive and the prescriptive in morality is complex and subtle. Contextualist morality strives for a nuanced understanding of the practice of morality that can serve as a basis for improving that morality.[15] Positivist morality so subordinates the descriptive that its "justification" becomes justification in the abstract, which ultimately is no justification at all. Morality can be critical only to the extent that it is informed, and the moral intelligibility that is the precursor to moral criticism needs the help of the social sciences.

NOTES

1. What I call "positivist morality" remains the dominant conception of morality within philosophy, and the recent proliferation of work in so-called "applied" ethics reflects this conception, as the name "applied ethics" itself demonstrates. Morality is taken to be a consistent, rationally justified system of norms that are applied to the facts of problematic cases to generate resolutions. Textbooks in medical ethics invariably are organized according to this positivist model. The first chapter is a survey of standard philosophical moral theories, such as Kantianism and utilitarianism, and the implication, conveyed overtly or covertly, is that the issues in ensuing chapters are to be addressed by applying these theories to them. Only recently, as a result of grappling with real-life moral problems, have philosophers begun to challenge the positivist view of morality and to develop alternative approaches to "practical" ethics.
2. This identification of the practice in an area with a theory about or a philosophy of that area has been rejected in the political realm by Walzer (1981) and Rorty (1988). In law Fish (1987: 1786) has criticized the view that "the better judge ... is the better philosopher."
3. For another set of cases that show how difficult it is to apply the notion of autonomy in clinical settings, see Jackson and Youngner (1979: 404-408).
4. This case is discussed by Corbin (1965: 164-170) and Michaels (1979: 23-34).
5. Urmson (1953: 38) recognized precisely this problem in his classic argument for interpreting Mill as a rule utilitarian: "in fact it is hard to state the relation of moral rules to a justifying principle with exactitude...."
6. Utilitarianism is the view that the right thing for an agent to do is adopt the course of action that is likely to produce the best consequences. Different versions of utilitarianism result, however, from difficulties in interpreting the notion of a course of action. Act utilitarianism assesses the consequences of individual, specific actions. Utilitarian generalization looks at the consequences of classes of action (in an attempt to capture the moral point of the question, "But what would happen if everybody did that?"). Rule utilitarianism considers the consequences of acting in accordance with a set of rules.
7. Horwitz (1982: 201-213) provides a clear, concise introduction to this problem.
8. Stone (1987) provides a provocative critique of what he calls "Moral Monism" and what I call positivist morality by challenging the assumption that only persons count for morality. Positivist morality remains insensitive to the crucial problem of ascertaining the domain of morality.
9. It might be thought that Daniels' notion of wide reflective equilibrium (WRE) solves this problem. Daniels claims, e.g., that "the detour of deriving the principles from the contract adds justificatory force to them, justification not found simply in the ... matching of principles and judgments" (1979: 261). As well, Daniels talks about the "intra-theory gain in justificatory force" (269) that

WRE provides. This putative justificatory force is the same as the third possibility for normativity in Rawls' view, namely, the revisability of considered judgments on the basis of principles. I do not have time to rebut this claim in detail, but in brief Daniels, too, ignores the distinction between occasions for and grounds of reassessment. Consequently, WRE permits "more drastic" revisions of considered judgments only in the sense that the occasions for such reassessments will be more frequent because the theory is more complex and thus there are more elements that can conflict with one another. But the ground for resolving such conflicts still remains considered judgments and not principles. In addition, Rawls seems to have abandoned the concern with justification. His recent work (Rawls 1980) suggests that he has adopted a contextualist approach to morality.

10. Rorty (1980: 39) makes this point. One of the most concrete and accurate accounts of how this proceeds comes from an anthropologist: "in everyday life the wise guide themselves as often by waiting to see what happens as by using rules, plans, and expectations. When in doubt, people find out about their worlds by living with ambiguity, uncertainty, or a simple lack of knowledge until the day, if and when it arrives, that their life experiences clarify matters. One can, in other words, learn by doing. Human beings often do muddle through, improvise, and make things up as they go along" (Rosaldo 1985: 20).

11. Fuller (1978) defends this position in law.

12. As Dreyfus (1980: 17) points out, exactly the same thing happens in the social sciences: "If, for the sake of agreement and 'objectivity,' the human sciences simply ignore their nonformalizable background practices, these practices will show up in each particular human science in the form of competing schools which call attention to what has been left out."

13. Comparative studies of how decisions are made and policies are formulated on moral matters could be particularly helpful. Glendon (1987) provides a good example.

14. Taylor (1971) provides an excellent account of the failings and limitations of a positivist approach to the social sciences.

15. In this vein one might wonder what a "clinical ethicist" working within a contextualist approach does. Because there is no single method of morality, there is no single answer to this question. Joffe's reproductive counselors provide one example of what a clinical ethicist could do in comparable contexts. In general, a clinical ethicist in the contextualist approach can perform all the jobs usually attributed to this role: be an analyst, an advocate, an advisor, an arbitrator, and an educator, for example. Baker (1988) provides a nice discussion of these functions as well as some others. And in general, which functions are appropriate and how they ought to be fulfilled will depend upon the context. The difference is that the performance of these functions will be informed and guided by an existing moral tradition not a philosophical moral theory.

REFERENCES

Baker, Robert
    1988   The Skeptical Critique of Clinical Ethics. *In* B. Hoffmaster, B. Freedman, and G. Fraser
            (eds.) Clinical Ethics in Theory and Practice. Clifton, N.J.: The Humana Press. Pp. 27-57.
Bayles, Michael
    1984   Moral Theory and Application. Social Theory and Practice 10: 97-120.
    1986   Mid-Level Principles and Justification. *In* Nomos XXVIII: Justification. J. Roland
            Pennock and John W. Chapman (eds.) New York: New York University Press. Pp. 49-67.
Corbin, Arthur L.
    1965   The Interpretation of Words and the Parol Evidence Rule. Cornell Law Quarterly 50:
            161-190.
Daniels, Norman
    1979   Wide Reflective Equilibrium and Theory Acceptance in Ethics. Journal of Philosophy 76:
            256-282.
Dreyfus, Hubert L.
    1980   Holism and Hermeneutics. Review of Metaphysics 34: 3-23.

Fish, Stanley
1987 Dennis Martinez and the Uses of Theory. Yale Law Journal 96: 1773-1800.
Fuller, Lon
1978 The Forms and Limits of Adjudication. Harvard Law Review 92: 353-409.
Geuss, Raymond
1981 The Idea of a Critical Theory. Cambridge, U.K.: Cambridge University Press.
Glendon, Mary Ann
1987 Abortion and Divorce in Western Law. Cambridge, Mass.: Harvard University Press.
Hart, H.L.A.
1961 The Concept of Law. Oxford: Oxford University Press.
Holmes, Oliver Wendell
1881 The Common Law. Boston: Little, Brown and Company.
Horwitz, Morton J.
1982 The Doctrine of Objective Causation. In David Kairys (ed.) The Politics of Law. New York: Pantheon Books. Pp. 201-213.
Jackson, David and Stuart Youngner
1979 Patient Autonomy and "Death with Dignity." New England Journal of Medicine 301, 8: 404-408.
Joffe, Carole
1986 The Regulation of Sexuality. Philadelphia: Temple University Press.
Jonsen, Albert
1986 Casuistry and Clinical Ethics. Theoretical Medicine 7: 65-74.
Lippman-Hand, Abby and F. Clarke Fraser
1979 Genetic Counseling: Parents' Responses to Uncertainty. Birth Defects: Original Article Series 15: 325-339.
Lyons, David
1965 Forms and Limits of Utilitarianism. New York: Oxford University Press.
Michaels, Walter Benn
1979 Against Formalism: The Autonomous Text in Legal and Literary Interpretation. Poetics Today 1: 23-34.
Miller, Bruce
1981 Autonomy and the Refusal of Lifesaving Treatment. Hastings Center Report 11, 4: 22-28.
Nell, Onora
1975 Acting on Principle. New York: Columbia University Press.
Noble, Cheryl
1979 Normative Ethical Theories. The Monist 62: 496-509.
O'Neill, Onora
1985 Consistency in Action. In Nelson Potter and Mark Timmons (eds.) Morality and Universality. Dordrecht: D. Reidel. Pp. 159-186.
Prichard, H.A.
1912 Does Moral Philosophy Rest on a Mistake? Mind 21: 21-37. Reprinted 1968. In H.A. Prichard. Moral Obligation. New York: Oxford University Press. Pp. 1-17.
Rawls, John
1951 Outline of a Decision Procedure for Ethics. Philosophical Review 60: 177-197.
1980 Kantian Constructivism in Moral Theory. Journal of Philosophy 77: 515-572.
Rorty, Richard
1980 A Reply to Dreyfus and Taylor. Review of Metaphysics 34: 39-46.
1988 The Priority of Democracy to Philosophy. In Merrill D. Peterson and Robert Vaugham (eds.) The Virginia Statute for Religious Freedom. Cambridge, U.K.: Cambridge University Press.
Rosaldo, Renato
1985 While Making Other Plans. Southern California Law Review 58: 19-28.

Stone, Christopher D.
    1987   Earth and Other Ethics. New York: Harper & Row.
Taylor, Charles
    1971   Interpretation and the Sciences of Man. Review of Metaphysics 25: 3-51.
Urmson, J.O.
    1953   The Interpretation of the Moral Philosophy of J.S. Mill. Philosophical Quarterly 3: 33-39.
           Reprinted 1968. *In* Michael Bayles (ed.) Contemporary Utilitarianism. Garden City, N.Y.:
           Doubleday and Company. Pp. 13-24.
Walzer, Michael
    1981   Philosophy and Democracy. Political Theory 9: 379-399.
    1985   Interpretation and Social Criticism. Cambridge, Mass.: Harvard University Press.

BRUCE JENNINGS

ETHICS AND ETHNOGRAPHY IN NEONATAL
INTENSIVE CARE

It is becoming increasingly difficult to find diehards who will flatly assert that bioethics and social science have nothing to contribute to (and learn from) one another. The rigid separation of facts and values once enshrined in academic discourse by the influence of logical positivism and Kantian formalism is now untenable. Fluid boundaries and blurred genres are now the order of the day throughout the humanities and the social sciences. Bioethics – as well as other areas of "applied" or "practical" ethics – operates at the heart of this ongoing reconfiguration of knowledge. In bioethics, normative and descriptive inquiry necessarily command equal time and attention, although this dual focus is naturally quite difficult to sustain.

As ethics by definition involves human relations and human doings, all work in ethics makes nearly constant reference to behavioral, psychological, and social description. However, philosophers will sometimes use descriptive materials in their work only for heuristic purposes or to illustrate something about the logic of moral concepts, much as, for example, Thomas Hobbes used his "descriptive" account of life in the state of nature as a vehicle for gaining insight into political power and authority. For such purposes the description need only be reasonable and plausible; it does not have to derive from some more systematic, rigorous process of social investigation. In bioethics, though, the "emic" experience of the philosopher is not sufficient because descriptive knowledge is not used simply for analytic purposes. The point of bioethics is to understand what action and practice should be, what is wrong with them now, why problems exist, and what can be done about those problems. None of these four aspects can be dropped without loss of something that is central to bioethics. And just as these questions interlock, so too do facts and values, descriptive and normative concepts, intertwine in virtually all corners of bioethics today.

Similarly, it is my sense (here I must speak less confidently) that the sociology and anthropology of medicine are becoming a good deal less positivistic than they once were. These studies are often based on a self-consciously hermeneutical approach to social action; that is to say, they give ideals, values, and norms, as well as notions of identity, purpose, and meaning, a serious place in their descriptive and explanatory accounts of action and practice. Such explanations rely on interpreting critically, and not simply reporting, the normative discourses constitutive of social practices. And this, in turn, requires conceptual (philosophical) as well as empirical methods and tools of analysis.

So the basic goals of bioethical and social scientific inquiry – together with a growing sophistication about matters of metatheory – should combine to bridge the gap between bioethics and the social scientific study of medicine (and the gap

G. Weisz (ed.), Social Science Perspectives on Medical Ethics, 261–272.
© 1990 Kluwer Academic Publishers. Printed in the Netherlands.

between applied ethics and interpretive social science generally). It is not that
bioethics supplies the normative ingredients, while social science supplies the
descriptive ingredients for some kind of hybrid. The point is that normative
inquiry and descriptive inquiry should have a place in *both* bioethics and the
social science of medicine (Jennings 1986).

Unhappily, the convergence that these considerations would lead one to expect
is not taking place; at least, not very often and not very quickly. Most writings in
bioethics still draw faintheartedly and unimaginatively, or not at all, on social
scientific studies that address the setting, institutional context, and cultural forces
relating to the bioethical problem at hand. All the while, of course, these writings
typically make large claims – or worse, assumptions – about setting, context, and
culture. These claims, in turn, are not harmless asides, sociological *obiter dicta*;
they do affect the normative ethical position offered in various ways. For one
thing, these sociological claims affect the reasonableness and persuasiveness of
the arguments made, since the reasonableness of an ethical argument that
balances conflicting values and weighs benefits and burdens often depends on the
context and setting within which this balancing and weighing goes on. Moreover,
background social claims and assumptions can quietly set the agenda of ethical
argument in the first place; they affect the very selection of problems and topics
to be discussed.

For their part, many ethnographic studies show little interest or subtlety in
analysing the normative languages spoken by their subjects. This failing is easy to
overlook until one runs across a work that does take this task seriously; then one
sees how rich the rewards can be. This is not an issue of adding judgment on top
of description; it pertains to the adequacy of description itself. When a social
scientist is unequipped to comprehend the conceptual and philosophical perplexities embedded in the form of life he or she is trying to describe and understand,
how adequate can that description be?

The interesting issue, then, is not whether there should be a rapprochement
between bioethics and social science, especially that variant we might generically
call "medical ethnography." The issue is why hasn't recent scholarship in these
two areas advanced further and faster along the road to more reciprocal
enlightenment? What obstacles impede such progress (for that is what I think it
would be)? How might we do better?

Part of the answer, I suppose, has to do with attitudes, education, and
temperament – philosophers and social scientists are not trained or inclined to
read, appreciate, or communicate with one another. With considerable effort, and
more opportunities for interdisciplinary collaboration these obstacles can be
overcome. Yet the deeper problem will remain until we develop richer, more
imaginative ways of formulating the connections between descriptive and
normative inquiry. In this essay, I propose to take a fresh look at these
connections, and to use the intersection of ethics and ethnography in neonatal
intensive care medicine in the United States as my vehicle for doing so.

I

Newborn intensive care is a particularly good subject for exploring the general difficulties – as well as the potential benefits – of forging a closer connection between ethnography and ethics for several reasons.

In the first place, the ethical and ethnographic literature on NICU care is now well enough developed and sufficiently overlapping for us to see what happens when ethnography and ethics (almost) get together. I will not attempt to survey all the literature here, by any means, but even a selective reading of a few recent book-length studies suggests a diagnosis of why the conversation never quite gets started. At any rate, the overlap between ethics and ethnography in studies of neonatal care is clear. Serious ethnographies of level III neonatal units (Anspach 1982; Frohock 1986; Guillemin and Holmstrom 1986; and Levin 1986), as well as more journalistic accounts (Gustaitis and Young 1986; and Lyon 1985), are deeply concerned with ethical and value issues. And significant new works on neonatal bioethics (Caplan et al. 1987; Kuhse and Singer 1985; Shelp 1986; and Weir 1984) are no less concerned with the cultural, technological, and institutional forces at work in the NICU.

Moreover, neonatal intensive care provides a good case study for our purposes because so much of the ethical discussion has focused on two key issues: (a) the proper process for making decisions about life-sustaining treatment with critically ill and severely impaired newborns; and (b) the use of the "best interest" standard and the nature of quality of life judgments that are made by providers and parents involved in the decisions. On both of these points, ethnographic studies of life within NICUS should provide telling information about how decisions are made and why. These studies should also shed light on the kinds of reforms that would be necessary if ethically preferred decision-making processes (e.g. those that give more power and responsibility to parents) were to be adopted in the future.

Finally, ethnography can contribute significantly to a well rounded ethical assessment by alerting us to some of the subtle costs and cultural consequences of reform efforts. Forms of medical practice and institutional patterns constitute an ecology of meanings where, as the maxim goes, you can't do just *one* thing. When ethics reforms misfire, the road to hell is paved with good intentions and ethnographic ignorance.

II

The recent ethnographic studies of NICUS usually describe several general characteristics of these settings that have a special significance for the bioethics of newborn care. The tertiary or level III NICU represents an unusually insular and self-contained institutional system, even within the world of specialized acute care in large medical centers. NICUS tend to be physically isolated from other hospital areas and access is limited for the protection of the infants. Although clear authority patterns exist, various NICU personnel work closely and intensively together, as neonatology is a "high touch" area of medicine for the physicians as well as for the nurses. Since a fairly large number of cases present some dramatic and ethically perplexing decision problems, a sub-culture of unit solidarity as well as decision-making by consensus tend to emerge. Moreover, NICUS seem, as cultural systems, to avoid using an explicitly ethical vocabulary when decisions are made; the "medicalization" of ethical questions is widespread.

A virtually overwhelming technological imperative operates on the entire ethos of the NICU, particularly on the physicians. Survival to discharge from the unit is the main focus and measure of success. Death is viewed not as a natural event but as a failure of medical practice. The most aggressive forms of treatment are applied on all but the most severely neurologically compromised infants. Follow-up after discharge from the unit is rare and, when it occurs, is usually selectively attentive to the best outcome cases.

American neonatal practice operates on the basis of treatment and evaluation "defaults" rather than by an ongoing evaluation and reassessment of individual patients over time (Rhoden 1986). Only a few categories of infants who reach the units are immediately identified for palliative care only. The prevailing moral calculus is to err on the side of over-treating some so that the moral risk of undertreating any will be minimized. Eventually recognized errors of overtreatment are readily forgiven.

Nurses tend to be less aggressive in their treatment judgments than physicians, but their role in case planning and management is limited. A single dissenting nurse's voice may override an impending decision not to treat aggressively, but it would never work the other way around. NICU nurses have much more "humanizing" contact with the infants (stroking, cuddling, rocking, talking) than do physicians.

The role of parents in the decision-making process is extremely limited once the infant is referred to the NICU. Geographical distance often makes it difficult for the parents to be in constant attendance, and in any case the emotional state of most parents makes them defer to the neonatologists, particularly in the early stages of treatment. Most of the daily or routine communication with parents takes place via the nurses. The only time the parent's voice is likely to be a significant factor is when aggressive treatment is being requested or when the neonatologists have come to their own independent conclusion that the baby is no longer "viable."

Despite the publicity and public interest it has generated in recent years, neonatal intensive care seems to be relatively unaffected by broader social forces and pressures. Cost-containment considerations do not yet seem to be a major factor in clinical decision-making. The effect of federal and state child protection regulations seems only to have reinforced an already very aggressive treatment posture prevalent in most units.

III

Paralleling these general findings of NICU ethnography are several basic points that have emerged in the recent bioethics literature on newborn care. These points are not universally accepted by all commentators, to be sure, but they do represent what can fairly be described as the mainstream position.

First, decisions to forgo life-sustaining treatment for neonates do not present *sui generis* ethical issues; they can be subsumed under the broader framework of principles and guidelines that govern life-sustaining treatment decisions for all patients. The basic principle of this framework is that no form of treatment is inherently mandatory; each treatment option should be weighed in terms of the burdens and benefits it offers the patient, and any form of treatment that is unduly burdensome to the patient may be forgone.

Having maintained that neonatal care does not require a radically distinct ethical framework, the bioethical literature nonetheless recognizes that all neonatal cases necessarily involve surrogate decision-making, and the special standards appropriate to it. Moreover, unlike cases of surrogate decision-making involving adults or even older children, it is impossible with newborns to make any reasonable inference concerning what the individual infant's values or treatment preferences might have been. The newborn has no personal history, no character, personality, or biographical relationships on which to base such an inference. Therefore, the only standard for surrogate decision-making that can be used is the so-called reasonable person standard or best interest standard. The precise content of "best interests" in these cases is what most of the ethical controversy is about, and it is made more difficult by the uncertainty that often surrounds the neonatal prognosis.

Whatever the content given to the best interests standard, there is widespread agreement that it should be interpreted as an infant-centered standard. That is, it is the best interests of the child, and not those of third parties, such as the infant's family or society more generally, that should count. Invidious judgments of "social worth" and social bias against those with a disability have no place in a best interests assessment. This is in keeping with the generally individualistic cast of the entire bioethics framework, where individual rights and equal respect for persons are paramount ethical orientations.

Finally, while no single procedural arrangement for making termination of treatment decisions will be perfect or foolproof, the best starting point is the presumption that the parents are the most trustworthy decision-makers for their child. Generally, parental authority and responsibility should be respected in preference to that of physicians, or the courts. Parental wishes should be challenged, reviewed, and overridden only when they are clearly beyond the pale of a reasonable judgment of the infant's best interests; simple value disagreements between parents and physicians should not be a sufficient basis on which to challenge parents' wishes.

IV

Setting the picture of neonatal care offered by the ethnographic literature side by side with that offered by the bioethics literature leads one to wonder whether ethnographers and ethicists are talking about the same institutional world. The stock answer that we are all used to is: no, they are not. The ethnographers are talking about the world of neonatology as it is, and the ethicists are talking about neonatology as it should be. That pat answer is, in my estimation, altogether misleading and inadequate.

There are several things that make the discrepancies between these two summaries less pronounced than they seem. Both sides, ethicists and ethnographers alike, are working on the basis of relatively limited fieldwork. Most ethicists (at least those who have written major works on neonatal care ethics) have done something of their own "fieldwork" by spending considerable time in NICUS and talking with the professionals who work there. This may not have been as extensive or as systematic as the fieldwork done by social scientists (Guillemin and Holmstrom have done the most extensive and intensive studies to date), but it does provide them with observations and anecdotes from which to generalize. So it is not correct to say that ethicists are simply talking about the way the NICU ought to be. They are often talking about the way they have discovered some NICUS to be at least some of the time, and they are generalizing from what they take to be real possibilities.

Contrary findings by ethnographers, which are also based on a limited sample, usually of one intensively studied NICU, can call the ethicists' expectations into question, to be sure. But it would be premature, at the very least, to conclude that the ethnographers have a "truer" grasp of the real world of the NICU than do the ethicists. The point here simply is that ethicists, if they are going to do something like ethnographic fieldwork anyway, ought to learn to do it well, and build it more explicitly into their research and writing. In that way we might begin to accumulate some findings that could more readily be combined and compared.

Of course, different sites and different ways of talking with informants and asking questions account for only some of the differences between the picture of the NICU presented by ethnographers and that offered by ethicists. And here the difference between what each side takes to be the *possibilities* of the world of the NICU is as important as the difference between what each side takes to be the actuality of the NICU. For it is not simply the ethicist who is talking about what the NICU could or should be like; the ethnographer, willy nilly, must have at least an implicit vision of that possibility as well.

This, then, is the second reason why it is misleading to conclude that the ethicist is simply talking about the ideal and the ethnographer the real. We have some capacity to determine the shape and conduct of neonatal intensive care in our system. What possibilities for change lie open to us, given where we are now? What direction should we take?

These, I suggest, are questions that both ethnography and ethics try to answer. And they are questions that neither in isolation has answered or can answer very well. Ethicists, by and large, are committed to rational persuasion and dialogue in order to sensitize neonatal professionals to the values at stake in treatment decision-making and care management. In the ethicists' vision of the ideal NICU, the twin evils of overtreatment and undertreatment would both be avoided, and caregivers would deal with prognostic uncertainty endemic to most NICU cases in an early stage by flexibly and progressively shifting from aggressive life-sustaining treatments (early on) to a more conservative strategy later. This would amount to individualized, rather than categorical, care planning in which weighing benefits and burdens and assessing quality of life would be openly and explicitly addressed. Once such value choices were acknowledged and made visible, neonatologists would presumably become more willing to involve parents in the decision-making process since medical experience and expertise alone are not sufficient to determine the infant's "best interest."

The ethnographers' vision of the ideal NICU is more nuanced and harder to glean from their analyses, since the conventions of ethnographic writing do not encourage the overt thematization of ethical reforms. Moreover, the stance of the ethnographer tends to be more detached than that of the ethicist; the ethnographer claims to observe without disturbing the form of life observed, and sometimes to disturb (the reader) simply by the force of observation. This, however, is a highly complex and value laden rhetorical posture in its own right (Geertz 1988). Like the neonatal ethicists, the neonatal ethnographers recognize the importance of shared communication among professionals and parents. The ethnographers also acknowledge the importance of individualized, flexible care planning. But they are usually less sanguine about the possibilities for change because of their perception of structural constraints on behavior.

Furthermore, in the ideal picture of the NICU that seems to be implicit in the ethnographic literature, the fabric holding the unit together is as important as the mode of treatment a particular infant receives. The values emphasized by ethnographers tend to be such things as shared meaning, coherence, and stability, all operating on several levels at once – the institutional level, the interpersonal level among staff, and within the personality of each caregiver. The naked reality of the conditions afflicting catastrophically ill newborns and the suffering that neonatal treatments often inflict are experiences that neonatal professionals must protect themselves from by means of special rituals, symbolism, and language. (If a decision is made to treat an extremely premature infant, it is a "baby"; if not, it is a "fetus.") Ethnographers study these aspects of life in the NICU as vehicles for expressing meaningful identity and sense of place in a complex human enterprise. Units in which interpersonal relations are "pathological" are implicitly contrasted with situations in which rituals and symbolism perform their social functions adequately.

V

Thus far I have argued that there are good metatheoretical reasons for thinking that ethnography and ethics ought to be complementary rather than conflicting modes of inquiry, but that the fields have been slow to seize that opportunity. I have also argued that it is superficial and misleading to construe the relationship between ethnography and ethics narrowly in terms of some purported complementarity (or conflict) between normative and descriptive inquiry. The fault lines between ethnography and ethics lie on a different and deeper level than that.

Lastly, in sketching some of the principal findings and positions in the recent ethnography and ethics of neonatology, I have suggested that these two fields right now seem to envision quite different possible future shapes for neonatal intensive care. Some of these differences might be accounted for by practice variations in the units that different researchers have studied or are most familiar with. But, at the end of the day, we must take these differences seriously and ask why ethnographers and ethicists see such contrasting things when they contemplate the prospects and possibilities for change.

There are at least three distinct reasons why ethnographers and ethicists tend to see social action and practice differently, and each of these is exemplified to some degree in the literature on neonatal intensive care.

The first is the relative degree of emphasis that is given to the agent, on the one hand, and to the context of agency, on the other. A different way to formulate the point is to say that ethics and ethnography can identify different things as the most significant unit of analysis. For ethics the individual is always the primary unit of analysis, and the agent, rather than the context of agency, is the main focus. Ethics looks at human choice and decisions made by individuals who are capable of informed, reasonable, and responsible deliberation. Ethics, for the most part, looks at the world-shaping power of human agency and at the exercise of that power by individual agents.

By contrast, ethnography generally tends to look at the action and agent-shaping power of the world. It highlights the context of agency, whether that context is conceptualized primarily in symbolic-cultural or institutional-social systemic terms. Indeed, contextualism can go so far that the individual human agent ceases to be a significant unit of analysis for explanatory purposes, and the "action" in question is not that of persons but of collectivities, structures, historical processes, and now even "discourses" or "grammars."

Now, as far as I can see, most of the recent ethnography on neonatal intensive care is relatively eclectic in theoretical orientation. Most studies provide a blend of role theory, symbolic interactionism, ethnomethodology – all of which, I take it, are theoretical orientations that do not actually push contextualism so far that a radical break with the agent-centered ethics orientation needs to be made. But it is not hard to imagine what, say, a Foucault-inspired study of the NICU would look like, or to see how it would intensify the tension with ethics far more than the American ethnography has.

Even in the American ethnographical studies, there are tendencies toward more radical contextualism. At various points, for instance, Guillemin and Holmstrom seem to regard NICUS as virtually uncontrollable institutional structures where the intake and technological management of critically ill infants has reached a kind of irresistible momentum and effectiveness. The NICU is nearly a self-perpetuating institutional system: it reproduces in the professionals who operate it the conditions for its own perpetuation and expansion (1986: 23-66). But Guillemin and Holmstrom do not pursue this theme consistently or develop it fully. Instead their concluding policy recommendations indicate a guarded optimism about the prospect of making changes that will curb the technological imperative currently at work in NICUS.

The second challenge that ethnography can pose to ethics is through what might be called the moral phenomenology developed by the ethnographic account. Just as ethics focuses on the foreground of individual agency and action rather than the background of context, it also takes as given the capacity to perceive issues as moral issues and to understand and use moral concepts and categories. I do not mean to suggest that ethicists overlook the fact that the ability to perceive and analyse ethical problems must be learned and practiced. Or that ethicists deny that our moral vocabulary grows out of certain traditions, is often ambiguous, and requires ongoing work to achieve clarity and definition. These matters, particularly the clarification of usage, consistency, and definition, certainly do concern ethicists, even to the point of preoccupying them. What ethicists do not do is see moral concepts and categories as embedded in ongoing forms of social practice and experience that are structured via particular institutional patterns or the encounter with certain technological constraints. And ethicists do not pay much attention to the ways in which struggling with a problem or acting within a certain pattern of constraints or power relationships can actually transform the moral perception and understanding of agents. Ethicists, like ethnographers, take moral vocabularies spoken by real people as social givens and starting points; few, if any, moral philosophers since Hobbes have tried to redefine the moral language wholesale. Ethicists tend to look at these vocabularies and languages in terms of their clarity, logical consistency, and scope. They do not look at how moral vocabularies can be functional or dysfunctional in particular settings and for specified purposes other than achieving moral clarity and "truth."

However, as philosophers as diverse as Stephen Toulmin, Alasdair MacIntyre, Bernard Williams, Stuart Hampshire, Michael Walzer, Martha Nussbaum, and Charles Taylor have shown, an understanding of the way moral notions are embedded in social practices – and of how moral perception, understanding and aspiration can be transformed by conduct within those practices and structures – is essential to the development of an adequate moral philosophy.

To shed light on these matters in a concrete and rich way is still a rare achievement in the social sciences, but ethnographic work is the venue where one might expect to find it. In Fred Frohock's *Special Care* (1986), the neonatal

literature gives us an especially fine example of this attention to moral phenomenology which does have significant implications for ethics. (Frohock, I suspect,
would object to my placing him in the "ethnographer's" camp, and indeed his
unconventional book is hard to pigeonhole.)

Frohock (and also Levin 1986) shows how much ethicists overlook when they
take notions like "treatment," "prognosis," "uncertainty," and "interests," let
alone "rights and "personhood," at face value. These are not basic terms with an
identifiable meaning that permits ethical debates over neonatal treatment decisions to be conducted. Instead these concepts are woven into and take their
meaning *in situ* from the medical nature of the newborn's condition, his or her
response to treatment, and the perceived personality traits and characteristics that
caregivers (especially nurses) read into the behavioral responses of the newborn
over time. (In a telling remark, Frohock points out that "[d]octors seem to bond
with disorders the way nurses bond with babies" [Frohock 1986: 93].) For the
purposes of ethical analysis, for example, newborns are "persons" simply by
virtue of their humanness; or they are "persons" if it is held that there are good
reasons to regard them as subjects of justice or bearers of rights. Frohock, on the
other hand, is acutely sensitive to the process through which neonatal professionals confer the status of personhood on the newborn. This focus is philosophically
important as well as descriptively interesting.

Frohock's work in the NICU reminds us that the moral status of the newborn is
not something that is simply given. It grows out of rational interaction in a
cultural setting that gives that interaction meaning. In ordinary settings, the moral
status of the newborn is not questioned and so may appear to be given. But in the
NICU the impairments of some newborns are so severe that the moral status of
personhood is much more apparently a badge of some kind of achievement. A
baby becomes a "person" because she is a "fighter," she exhibits a will to live, an
impulse to thrive. Just as personhood in this sense comes into being, it can also
recede and be extinguished. Neonatal professionals are quick to realize when a
baby has "given up" and has begun dying (Frohock 1986: 74-115). If they are not
careful, then, ethicists will miss the meanings that neonatal physicians and nurses
convey when they use the (deceptively familiar) medical-moral terminology that
we all use to discuss these kinds of cases. Just as bad, the ethicists will misjudge
how their own message and recommendations (conveyed in this vocabulary) will
be received and understood.

Notice that the problem here is *not* jargon, or the inability of each side to
understand the other's own special habits of definition and usage. The problem on
the medical side (but on the philosopher's side as well) is the relationship
between moral language and a cultural-institutional world. It is the proper
business of the ethnographer to understand that relationship, to bring it into the
light of day, and thereby to play a mediating role in the communication between
ethicists and neonatologists.

The third and final way in which ethnography can challenge ethics is more
familiar and can be mentioned here more briefly. Bioethics generally has a

simple, not to say simpleminded, notion of what can be done to bring about social change – in most cases to reform professional practice to bring it more into line with established ethical obligations and principled responsibilities. The strategy is: argument, agreement through rational persuasion, and education. The commitment to this polis model runs very deeply in philosophy, and the applied ethics movement of the past twenty years has been premised on the belief that it can be brought out of the confines of the academy and introduced into the conduct of public and professional life.

I do not think that social science has shown – or is likely to show – that this belief of the applied ethicist is false, or silly, or somehow unworthy. But what social scientific studies have done, and neonatal ethnography is particularly insistent in this regard, is to force ethicists to pay more attention to the cultural, institutional, and psychological preconditions for social and behavioral change.

In neonatal care this is especially important in relation to the question of the parents' role in treatment decision-making and planning. Almost all ethical opinion holds that the parents' rights and responsibilities are compelling; it is both wrong and harmful to disenfranchise them. At the same time, the clear message from studies of NICUS is that parents do not now play a very meaningful role. And without major changes in the way NICUS are organized, professionals are trained, and parents are counseled, it is unlikely that they ever will. If, aside from their concern for parents' rights, ethicists look to the parents as the best hope for controlling the technological imperative and countering the overwhelming bias in favor of overtreatment, then far more than rational persuasion is going to be necessary.

So far this particular ethnographic message has not gotten through to the ethicists. One still finds very elegant discussions of parental roles and the cooperative role of the doctor with no reference whatever to the actual structure and operations of the NICU (Cf. Shelp 1986: 77-107).

CONCLUSION

On the whole I think that the neonatal ethnographers have been more influenced by the neonatal ethicists than the other way around. As the ethnographic studies increase and improve (although the early work has set a high standard, it seems to me), that will change. Social science can improve bioethics, not just by providing better facts, but by showing bioethics, through an exploration of the moral life, the way to do better ethics.

REFERENCES

Anspach, Renée
    1982   Life and Death Decisions in Neonatal Intensive Care; A Study in the Sociology of Knowledge. University of California, San Diego: Ph.D. Dissertation.

Caplan, Arthur L. et al.
    1987   Imperiled Newborns. Hastings Center Report 17, 6: 5-32.
Frohock, Fred M.
    1986   Special Care: Medical Decisions at the Beginning of Life. Chicago: University of Chicago
            Press.
Geertz, Clifford
    1988   Works and Lives: The Anthropologist as Author. Stanford: Stanford University Press.
Guillemin, Jeanne Harley and Lynda Lytle Holmstrom
    1986   Mixed Blessings: Intensive Care for Newborns. New York: Oxford University Press.
Gustaitis, Rasa and Ernle W.D. Young
    1986   A Time to Be Born, A Time to Die: Conflicts and Ethics in an Intensive Care Nursery.
            Reading: Addison-Wesley.
Jennings, Bruce
    1986   Applied Ethics and the Vocation of Social Science. In Joseph DeMarco and Richard Fox
            (eds.) New Directions in Ethics. London: Routledge and Kegan Paul. Pp. 205-217.
Kuhse, Helga and Peter Singer
    1985   Should the Baby Live? New York: Oxford University Press.
Levin, Betty Wolder
    1986   Caring Choices: Decision Making About Treatment for Catastrophically Ill Newborns.
            Columbia University: Ph.D. Dissertation.
Lyon, Jeff
    1985   Playing God in the Nursery. New York: Norton.
Rhoden, Nancy
    1986   Treating Baby Doe: The Ethics of Uncertainty. Hastings Center Report 16, 4: 34-42.
Shelp, Earl E.
    1986   Born to Die? Deciding the Fate of Critically Ill Newborns. New York: The Free Press.
Weir, Robert
    1984   Selective Nontreatment of Handicapped Newborns. New York: Oxford University Press.

BOOK-LENGTH WORKS REFLECTING OR RELEVANT TO SOCIAL SCIENCE PERSPECTIVES ON MEDICAL ETHICS

Alderson, D.P.
   1988 Informed Consent: Problems of Parental Consent to Paediatric Cardiac Surgery. London University: Ph.D. Dissertation.
Aly, Götz (ed.)
   1987 Aktion T4, 1939-1945: Die "Euthanasie" Zentrale in der Tiergartenstrasse 4. West Berlin: Edition Hentrich.
Ambroselli, Claire
   1988 L'éthique médicale. Paris: Presses Universitaires de France.
Anspach, Renée
   1982 Life and Death Decisions in Neonatal Intensive Care: A Study in the Sociology of Knowledge. University of California, San Diego: Ph.D. Dissertation.
Apfel, Roberta J. and Susan M. Fisher
   1984 To Do No Harm: DES and the Dilemmas of Modern Medicine. New Haven: Yale University Press.
Arditti, Rita et al. (eds.)
   1984 Test-Tube Women: What Future for Motherhood? London: Pandora Press.
Barber, Bernard
   1980 Informed Consent in Medical Therapy and Research. New Brunswick, N.J.: Rutgers University Press.
Barber, Bernard et al.
   1973 Research on Human Subjects: Problems of Social Control in Medical Experimentation. New York: Russell Sage Foundation.
Barnes, J.A.
   1979 Who Should Know What? Social Science, Privacy and Ethics. Harmondsworth: Penguin.
Berlant, Jeffrey L.
   1975 Profession and Monopoly: A Study of Medicine in the United States and Great Britain. Berkeley: University of California Press.
Bluebond-Langner, Myra
   1978 The Private Worlds of Dying Children. Princeton, N.J.: Princeton University Press.
Bock, Gisela
   1986 Zwangssterilisation im Nationalsozialismus: Studien zur Rassenpolitik und Frauenpolitik. Opladen: Westdeutscher Verlag.

Bosk, Charles L.
    1979 Forgive and Remember: Managing Medical Failure. Chicago: Univer-
        sity of Chicago Press.
Brand, Ulrich
    1977 Aertzliche Ethik im 19. Jahrhundert; der Wandel ethischer Inhalte im
        medizinischen Schrifttum. Ein Beitrag zum Verstandnis der Arzt-
        Patient-Beziehung. Freiburg i. Br.: Hans Ferdinand Schulz Verlag.
Brookes, Barbara
    1988 Abortion in England 1900-1967. London: Croom Helm.
Burns, Chester Ray (ed.)
    1977 Legacies in Ethics and Medicine. New York: Science History Publica-
        tions.
Callahan, Daniel and Bruce Jennings (eds.)
    1983 Ethics, the Social Sciences, and Policy Analysis. New York: Plenum
        Press.
Carrick, Paul
    1985 Medical Ethics in Antiquity: Philosophical Perspectives on Abortion
        and Euthanasia. Dordrecht: D. Reidel.
Crane, Diana
    1975 The Sanctity of Social Life: Physicians' Treatment of Critically Ill
        Patients. New York: Russell Sage Foundation.
Culhane Speck, D.
    1987 An Error in Judgement: The Politics of Medical Care in An Indian/
        White Community. Vancouver: Talonbooks.
Dowie, J. and Arthur S. Elstein (eds.)
    1988 Professional Judgement. A Reader in Clinical Decision Making. Cam-
        bridge, U.K.: Cambridge University Press.
Edelstein, Ludwig
    1967 Ancient Medicine: Selected Papers. Baltimore: Johns Hopkins Univer-
        sity Press.
Eisenberg, Leon and Arthur Kleinman (eds.)
    1981 The Relevance of Social Science for Medicine. Dordrecht: D. Reidel.
Fox, Renée C.
    1959 Experiment Perilous: Physicians and Patients Facing the Unknown.
        Glencoe, Ill.: The Free Press.
    1989 The Sociology of Medicine: A Participant Observer's View. Englewood
        Cliffs, N.J.: Prentice-Hall.
Fox, Renée C. and Judith P. Swazey
    1978 The Courage to Fail: A Social View of Organ Transplants and Dialysis.
        (2nd ed.) Chicago: University of Chicago Press.
Freidson, Eliot
    1970 Profession of Medicine: A Study of the Sociology of Applied Knowl-
        edge. New York: Harper & Row.

1975 Doctoring Together: A Study of Professional Social Control. New York: Elsevier.

Frohock, Fred M.
1986 Special Care: Medical Decisions at the Beginning of Life. Chicago: University of Chicago Press.

Fuchs, Victor R.
1974 Who Shall Live? Health, Economics, and Social Choice. New York: Basic Books.

Glaser, Barney G. and Anselm L. Strauss
1965 Awareness of Dying. Chicago: Aldine.
1968 Time for Dying. Chicago: Aldine.

Glendon, Mary Ann
1987 Abortion and Divorce in Western Law: American Failures, European Challenges. Cambridge, Mass.: Harvard University Press.

Gray, Bradford H.
1975 Human Subjects in Medical Experimentation: A Sociological Study of the Conduct and Regulation of Clinical Research. New York: Wiley & Sons.

Gray, Bradford H. (ed.)
1986 For-Profit Enterprise in Health Care. Committee on Implications of For-Profit Enterprise in Health Care, Institute of Medicine. Washington, D.C.: National Academy Press.

Guillemin, Jeanne H. and Lynda L. Holstrom
1986 Mixed Blessings: Intensive Care for Newborns. New York: Oxford University Press.

Hahn, Robert A. and Atwood D. Gaines (eds.)
1985 Physicians of Western Medicine: Anthropological Approaches to Theory and Practice. Dordrecht: D. Reidel.

Harwood, Alan (ed.)
1981 Ethnicity and Medical Care. Cambridge, Mass.: Harvard University Press.

Hatch, Elvin
1983 Culture and Morality: The Relativity of Values in Anthropology. New York: Columbia University Press.

Imber, Jonathan B.
1986 Abortion and the Private Practice of Medicine. New Haven: Yale University Press.

Ingman, Stanley R. and Derek G. Gill (eds.)
1986 Geriatric Care and Distributive Justice. Special Issue of Social Science and Medicine 23, 12.

Isambert, François-André and Gwen Terrenoire
1986 Éthique des sciences de la vie et de la santé. Paris: La Documentation Française.

Joffe, Carol
   1986 The Regulation of Sexuality: Experiences of Family Planning Workers. Philadelphia: Temple University Press.
Jones, James H.
   1981 Bad Blood. The Tuskegee Syphilis Experiment. New York: The Free Press.
Keown, John
   1988 Abortion, Doctors, and the Law: Some Aspects of the Legal Regulation of Abortion in England from 1803 to 1982. Cambridge, U.K.: Cambridge University Press.
Kevles, Daniel
   1985 In the Name of Eugenics. New York: Alfred A. Knopf.
Klee, Ernst
   1985 "Euthanasie" im NS-Staat: Die "Vernichtung lebensunwerten Lebens." Frankfurt am Main: S. Fischer.
Kleinman, Arthur
   1988 The Illness Narratives: Suffering, Healing, and the Human Condition. New York: Basic Books.
Koenig, Barbara A.
   1987 The Technological Imperative in Medical Practice: An Ethnographic Study of Therapeutic Plasma Exchange. University of California, San Diego: Ph.D. Dissertation.
Konold, Donald E.
   1962 A History of American Medical Ethics, 1847-1912. Madison, Wis.: State Historical Society of Wisconsin.
Korbin, Jill (ed.)
   1981 Child Abuse and Neglect: Cross-Cultural Perspectives. Berkeley: University of California Press
Levey, Martin
   1967 Medical Ethics of Medieval Islam with Special Reference to al-Ruhawi's *Practical Ethics of the Physician*. Transactions of the American Philosophical Society 57, 3.
Levin, Betty Wolder
   1986 Caring Choices: Decision Making About Treatment for Catastrophically Ill Newborns. Columbia University: Ph.D. Dissertation.
Lewin, Ellen and Virginia Olesen (eds.)
   1985 Women, Health and Healing: Toward a New Perspective. New York: Tavistock.
Lidz, Charles W. et al.
   1984 Informed Consent: A Study of Decisionmaking in Psychiatry. New York: The Guilford Press.
Light, Donald
   1980 Becoming Psychiatrists: The Professional Transformation of Self. New York: W.W. Norton & Co.

Light, Donald W. and Alexander Schuller (eds.)
    1986 Political Values and Health Care: The German Experience. Cambridge,
        Mass.: The MIT Press.
Lock, Margaret and Deborah Gordon (eds.)
    1988 Biomedicine Examined. Dordrecht: Kluwer Academic Publishers.
Luker, Kristin
    1984 Abortion and the Politics of Motherhood. Berkeley: University of
        California Press.
Marks, Harry Milton
    1987 Ideas as Reforms: Therapeutic Experiments and Medical Practice,
        1900-1980. Massachusetts Institute of Technology: Ph.D. Dissertation.
McKinlay, John B.
    1982 Technology and the Future of Health Care. Cambridge, Mass.: The MIT
        Press.
McLaren, Angus and Arlene Tigar McLaren
    1986 The Bedroom and the State: The Changing Practices and Politics of
        Contraception and Abortion in Canada, 1880-1980. Toronto: Mc-
        Clelland & Stewart.
Mechanic, David
    1979 Future Issues in Health Care: Social Policy and the Rationing of
        Medical Services. New York: The Free Press.
    1986 From Advocacy to Allocation: The Evolving American Health Care
        System. New York: The Free Press.
Mizrahi, Terry
    1986 Getting Rid of Patients: Contradictions in the Socialization of Physi-
        cians. New Brunswick, N.J.: Rutgers University Press.
Mohr, James C.
    1978 Abortion in America: The Origins and Evolution of National Policy,
        1800-1900. New York: Oxford University Press.
Nowak, Kurt
    1980 "Euthanasie" und Sterilisierung im "Dritten Reich": Die Konfrontation
        der evangelischen und katcholischen Kirche mit dem "Gesetz zur
        Verhütung erbkranken Nachwuchses" und der "Euthanasie"-Aktion.
        Göttingen: Vandenhoeck & Ruprecht.
O'Brien, Mary Elizabeth
    1983 The Courage to Survive: The Life Career of the Chronic Dialysis
        Patient. New York: Grune and Stratton.
Pernick, Martin S.
    1985 A Calculus of Suffering: Pain, Professionalism, and Anesthesia in
        Nineteenth-Century America. New York: Columbia University Press.
Petchesky, Rosalind Pollack
    1984 Abortion and Woman's Choice: The State, Sexuality and Reproductive
        Freedom. New York: Longman.

Plough, Alonzo L.
    1986 Borrowed Time: Artificial Organs and the Politics of Extending Lives. Philadelphia: Temple University Press.
Proctor, Robert N.
    1988 Racial Hygiene: Medicine Under the Nazis. Cambridge, Mass.: Harvard University Press.
Reed, James
    1978 From Private Vice to Public Virtue: The Birth Control Movement and American Society Since 1830. New York: Basic Books.
Reiser, Stanley J.
    1978 Medicine and the Reign of Technology. Cambridge, U.K.: Cambridge University Press.
Richter, Gabriel
    1986 Blindheit und Eugenik 1918-1945. Freiburg i. Br.: H.F. Schulz.
Scheper-Hughes, Nancy (ed.)
    1987 Child Survival: Anthropological Perspectives on the Treatment and Maltreatment of Children. Dordrecht: D. Reidel.
Schneider, Carl E. and Maris A. Vinovskis
    1980 The Law and Politics of Abortion. Lexington, Mass.: D.C. Heath.
Scully, Diana
    1980 Men Who Control Women's Health: The Miseducation of Obstetrician-Gynecologists. Boston: Houghton Mifflin.
Searle, G.R.
    1976 Eugenics and Politics in Britain, 1900-1914. Leyden: Noordhoff International Publishers.
Simmons, Roberta G. et al.
    1977 Gift of Life: The Social and Psychological Impact of Organ Transplantation. New York: Wiley & Sons. (2nd edition published 1987, Transaction Books.)
Soloway, Richard
    1982 Birth Control and the Population Question in England, 1877-1930. Chapel Hill: University of North Carolina Press.
Starr, Paul
    1982 The Social Transformation of American Medicine. New York: Basic Books.
Steinhoff, Patricia G. and Milton Diamond
    1977 Abortion Politics: The Hawaii Experience. Honolulu: University Press of Hawaii.
Stimson, Gerry and Barbara Webb
    1975 Going to See the Doctor: The Consultation Process in General Practice. London: Routledge and Kegan Paul.
Strauss, Anselm et al.
    1985 Social Organization of Medical Work. Chicago: University of Chicago Press.

Strauss Anselm L. and Barney G. Glaser
  1970 Anguish: A Case History of A Dying Trajectory. Mill Valley, California: The Sociology Press.
Sudnow, David
  1967 Passing On: The Social Organization of Dying. Englewood Cliffs, N.J.: Prentice-Hall.
Taylor, Kathryn
  1984 Decision Difficult: Physician Behaviour in the Diagnosis and Treatment of Breast Cancer. McGill University: Ph.D. Dissertation.
Unschuld, Paul U.
  1979 Medical Ethics in Imperial China: A Study in Historical Anthropology. Berkeley: University of California Press.
Valenstein, Elliot S.
  1986 Great and Desperate Cures: The Rise and Decline of Psychosurgery and Other Radical Treatments for Mental Illness. New York: Basic Books.
Veatch, Robert M. (ed.)
  1988 Comparative Medical Ethics. Special Issue of Journal of Medicine and Philosophy 13.
Villey, Raymond
  1986 Histoire du secret médical. Paris: Editions Seghers.
Woycke, James
  1988 Birth Control in Germany. London: Routledge.
Zaner, Richard M. (ed.)
  1988 Death: Beyond Whole-Brain Criteria. Dordrecht: D. Reidel.

SELECTED LIST OF BOOK-LENGTH WORKS ON OR BASIC TO MEDICAL ETHICS

Appelbaum, Paul S., Charles W. Lidz and Alan Meisel
  1987 Informed Consent: Legal Theory and Clinical Practice. New York: Oxford University Press.
Baumrin, Bernard and Benjamin Freedman
  1983 Moral Responsibility and the Professions. New York: Haven Publications.
Bayles, Michael
  1984 Reproductive Ethics. Englewood Cliffs, N.J.: Prentice-Hall.
Beauchamp, Tom L. and James F. Childress
  1983 Principles of Biomedical Ethics. (2nd ed.) New York: Oxford University Press.
Beauchamp, Tom L. and L.B. McCullough
  1984 Medical Ethics: The Moral Responsibilities of Physicians. Englewood Cliffs, N.J.: Prentice-Hall.
Benjamin, Martin and Joy Curtis
  1981 Ethics in Nursing. New York: Oxford University Press.
Brody, Howard
  1976 Ethical Decisions in Medicine. Boston: Little, Brown and Company.
Callahan, Daniel
  1987 Setting Limits: Medical Goals in an Aging Society. New York: Simon and Schuster.
Christie, Ronald and Barry Hoffmaster
  1986 Ethical Issues in Family Medicine. New York: Oxford University Press.
Daedalus
  1969 Special Issue: Ethical Aspects of Human Experimentation.
Daniels, Norman
  1985 Just Health Care: Studies in Philosophy and Health Policy. Cambridge, U.K.: Cambridge University Press.
DeMarco, Joseph P. and Richard M. Fox (eds.)
  1986 New Directions in Ethics: The Challenge of Applied Ethics. London: Routledge and Kegan Paul.
Devlin, Patrick
  1965 The Enforcement of Morals. Oxford: Oxford University Press.
Engelhardt, H. Tristram
  1986 The Foundations of Bioethics. New York: Oxford University Press.
Engelhardt, H. Tristram and Daniel Callahan (eds.)
  1977 Knowledge, Value and Belief. Hastings-on-Hudson, N.Y.: The Hastings Center.

Faden, R.R. and Beauchamp, T.L.
    1986 A History and Theory of Informed Consent. New York: Oxford University Press.
Fagot-Largault, Anne
    1985 L'homme bioéthique: pour une déontologie de la recherche sur le vivant. Paris: Maloine.
Fletcher, Joseph Francis
    1954 Morals and Medicine; The Moral Problems of: The Patient's Right to Know the Truth, Contraception, Artificial Insemination, Sterilization, Euthanasia. Princeton, N.J.: Princeton University Press.
    1966 Situation Ethics: The New Morality. Philadelphia: Westminster Press.
    1979 Humanhood: Essays in Biomedical Ethics. Buffalo: Prometheus Books.
Garfield, Jay L. and Patricia Hennessey (eds.)
    1984 Abortion: Moral and Legal Perspectives. Amherst, Mass.: University of Massachusetts Press.
Graber, Glenn C. et al.
    1985 Ethical Analysis of Clinical Medicine: A Guide to Self Evaluation. Baltimore: Urban and Schwarzenberg.
Hanen, Marsha and Kai Nielson (eds.)
    1987 Science, Morality and Feminist Theory. Canadian Journal of Philosophy, Supplementary Volume 13. Calgary: University of Calgary Press.
Harding, Sandra
    1986 The Science Question in Feminism. Ithaca: Cornell University Press.
Harris, John
    1985 The Value of Life: An Introduction to Medical Ethics. London: Routledge and Kegan Paul.
Hartmann, Betsy
    1987 Reproductive Rights and Wrongs: The Global Politics of Population Control and Contraceptive Choice. New York: Harper & Row.
Jakobovits, Immanuel
    1975 Jewish Medical Ethics: A Comparative and Historical Study of the Jewish Religious Attitude to Medicine and its Practice. (2nd ed.) New York: Bloch.
Katz, Jay
    1984 The Silent World of Doctor and Patient. New York: The Free Press.
Levine, Robert J.
    1986 Ethics and Regulation of Clinical Research. (2nd ed.) Baltimore: Urban and Schwarzenberg.
MacIntyre, Alasdair
    1981 After Virtue: A Study in Moral Theory. South Bend, Ind.: Notre Dame University Press.
Mappes, Thomas A. and Jane S. Zembaty (eds.)
    1981 Biomedical Ethics. New York: McGraw-Hill.

Menzel, Paul T.
  1983 Medical Costs, Moral Choices: A Philosophy of Health Care Economics in America. New Haven: Yale University Press.
Nell, Onora
  1975 Acting on Principle: An Essay on Kantian Ethics. New York: Columbia University Press.
Okinczyk, Joseph
  1936 Humanisme et médecine. Paris: Labergerie.
Ordre National des Médecins
  1955 Premier Congrès International de Morale Médicale, I, Rapports; II, Communications, Compte Rendu. Paris: Masson.
  1966 Deuxième Congrès International de Moral Médicale, I, Rapports; II, Communications, Compte Rendu. Paris: Ordre National des Médecins.
Pellegrino, Edmund D. and David C. Thomasina
  1981 A Philosophical Basis of Medical Practice: Toward a Philosophy and Ethic of the Healing Professions. New York: Oxford University Press.
Pellegrino, Edmund D. and Thomas K. McElhinney
  1981 Teaching Ethics, the Humanities, and Human Values in Medical Schools: A Ten-Year Overview. Washington, D.C.: Institute on Human Values in Medicine, Society for Health and Human Values.
Portes, Louis
  1954 A la recherche d'une éthique médicale. Paris: Masson and Presses Universitaires de France.
Potter, Nelson and Mark Timmons (eds.)
  1985 Morality and Universality: Essays on Ethical Universalizability. Dordrecht: D. Reidel.
Potter, Van Rensselaer
  1971 Bioethics: Bridge to the Future. Englewood Cliffs, N.J.: Prentice-Hall.
President's Commission for the Study of Ethical Problems in Medicine and Biomedical and Behavioral Research
  1982 Making Health Care Decisions: The Ethical and Legal Implications of Informed Consent in the Patient-Practitioner Relationship. Washington, D.C.: U.S. Government Printing Office.
  1983 Securing Access to Health Care. Volume One: Report. The Ethical Implications of Differences in the Availability of Health Services. Washington, D.C.: U.S. Government Printing Office.
Prichard, H.A.
  1949 Moral Obligation. Oxford: Clarendon Press.
Ramsey, Paul
  1970 The Patient as Person: Explorations in Medical Ethics (The Lyman Beecher Lectures). New Haven: Yale University Press.
Rawls, John
  1971 A Theory of Justice. Cambridge, Mass.: Harvard University Press.

Reiser, Stanley et al.
　　1977 Ethics in Medicine: Historical Perspectives and Contemporary Concerns. Cambridge, Mass.: The MIT Press.
Rosner, Fred and J. David Bleich (eds.)
　　1979 Jewish Bioethics. New York: Sanhedrin Press.
Schwartz, D., R. Flamant and J. Lellouch
　　1981 L'essai thérapeutique chez l'homme. Paris: Flammarion.
Shelp, Earl E. (ed.)
　　1985 Virtue and Medicine: Explorations in the Character of Medicine. Dordrecht: D. Reidel.
Shelp, Earl E.
　　1986 Born to Die? Deciding the Fate of Critically Ill Newborns. New York: The Free Press.
Singer, Peter and Helga Kuhse
　　1985 Should the Baby Live? The Problems of Handicapped Babies. New York: Oxford University Press.
Singhal, Ghanshyam Das and Damodar Sharma Gaur
　　1963 Surgical Ethics in Ayerveda. Varanasi, India: Chowkhamba Sanskrit Series Office.
Tancredi, Lawrence R. (ed.)
　　1974 Ethics of Health Care. Washington, D.C.: National Academy of Sciences.
Veatch, Robert
　　1976 Death, Dying and the Biological Revolution: Our Last Quest for Responsibility. New Haven: Yale University Press.
　　1981 A Theory of Medical Ethics. New York: Basic Books.
Weir, Robert
　　1984 Selective Nontreatment of Handicapped Newborns. New York: Oxford University Press.

AUTHOR/NAME INDEX

Italicized numbers refer to full bibliographic references.

285

SUBJECT INDEX

291

The Culture, Illness, and Healing Book Series

Editors:
Margaret Lock, *McGill University, Montreal, Canada*
Allan Young, *McGill University, Montreal, Canada*

Publications:
1. L. Eisenberg and A. Kleinman (eds.): *The Relevance of Social Science for Medicine.* 1981   ISBN Hb: 90–277–1176–3; Pb: 90–277–1185–2
2. A. Kleinman and Tsung-yi Lin (eds.): *Normal and Abnormal Behavior in Chinese Culture.* 1981   ISBN 90–277–1104–6
3. C. F. Sargent: *The Cultural Context of Therapeutic Choice.* Obstetrical Care Decisions Among the Bariba of Benin. 1982   ISBN 90–277–1344–8
4. A. J. Marsella and G. M. White (eds.): *Cultural Conceptions of Mental Health and Therapy.* 1982   ISBN Hb: 90–277–1362–6; Pb: 90–277–1757–5
5. N. J. Chrisman and Th. W. Maretzki (eds.): *Clinically Applied Anthropology.* Anthropologists in Health Science Settings. 1982
                    ISBN Hb: 90–277–1418–5; Pb: 90–277–1419–3
6. R. A. Hahn and A. D. Gaines (eds.): *Physicians of Western Medicine,* Anthropological Approaches to Theory and Practice. 1985
                    ISBN Hb: 90–277–1790–7; Pb: 90–277–1881–4
7. R. C. Simons and C. C. Hughes (eds.): *The Culture-Bound Syndromes.* Folk Illnesses of Psychiatric and Anthropological Interest. 1985
                    ISBN Hb: 90–277–1858–X; Pb: 90–277–1859–8
8. L. L. Langness and H. G. Levine (eds.): *Culture and Retardation.* Life Histories of Mildy Mentally Retarded Persons in American Society. 1986
                    ISBN Hb: 90–277–2177–7; Pb: 90–277–2178–5
9. C. R. Janes, R. Stall and S. M. Gifford (eds.): *Anthropology and Epidemiology.* Interdisciplinary Approaches to the Study of Health and Disease. 1986
                    ISBN Hb: 90–277–2248–X; Pb: 90–277–2249–8
10. J. G. Kennedy: *The Flower of Paradise.* The Institutionalized Use of the Drug Qat in North Yemen. 1987
                    ISBN Hb: 1–55608–011–5; Pb: 1–55608–012–3
11. N. Scheper-Hughes (ed.): *Child Survival.* Anthropological Perspectives on the Treatment and Maltreatment of Children. 1987
                    ISBN Hb: 1–55608–028–X; Pb: 1–55608–029–8
12. S. van der Geest and S. Reynolds Whyte (eds.): *The Context of Medicines in Developing Countries.* Studies in Pharmaceutical Anthropology. 1988
                    ISBN 1–55608–059–X
13. M. Lock and D. Gordon (eds.): *Biomedicine Examined.* 1988
                    ISBN Hb: 1–55608–071–9; Pb: 1–55608–072–7
14. S. Frankel and G. Lewis (eds.): *A Continuing Trial of Treatment.* Medical Pluralism in Papua New Guinea. 1989
                    ISBN Hb: 1–55608–076–X; Pb: 1–55608–078–5
15. M. Nichter: *Anthropology and International Health.* South Asian Case Studies. 1989   ISBN Hb: 1–55608–005–X; Pb: 0–7923–0158–7
16. G. Weisz (ed.): *Social Science Perspectives on Medical Ethics.* 1990
                    ISBN 0-7923-0566-3